THE INFLUENCE OF COMPUTING
ON MATHEMATICAL RESEARCH AND EDUCATION

PROCEEDINGS OF SYMPOSIA
IN APPLIED MATHEMATICS
Volume 20

THE INFLUENCE OF COMPUTING
ON MATHEMATICAL RESEARCH AND EDUCATION

AMERICAN MATHEMATICAL SOCIETY
PROVIDENCE, RHODE ISLAND
1974

PROCEEDINGS OF THE SYMPOSIUM IN APPLIED MATHEMATICS
OF THE AMERICAN MATHEMATICAL SOCIETY

HELD AT THE UNIVERSITY OF MONTANA
MISSOULA, MONTANA
AUGUST 14–17, 1973

EDITED BY
JOSEPH P. LASALLE

Prepared by the American Mathematical Society

with partial support from National Science Foundation grant GJ-37714

Library of Congress Cataloging in Publication Data

Symposium in Applied Mathematics, 20th, University of
  Montana, Missoula, 1973.
    The influence of computing on mathematical research
and education.

    (Proceedings of symposia in applied mathematics,
v. 20)
    Prepared by the American Mathematical Society.
    Bibliography: p.
    1. Electronic data processing—Mathematics—Con-
gresses. 2. Mathematical research—Congresses.
3. Mathematics—Study and teaching—Congresses.
I. LaSalle, Joseph P., ed. II. American Mathematical Society. III. Title. IV. Series.

QA1.S894 1973 510'.7 74-5166
ISBN 0-8212-1326-1

AMS (MOS) subject classifications.
00A10, 98A05, 98A35, 98B20,
10-04, 68A10, 68A40, 13-04, 20-04, 62-04, 60-04

CONTENTS

Indexes

Preface

This volume contains six of the invited addresses and fourteen of the contributed papers that were presented at the joint American Mathematical Society and the Mathematical Association of America Conference on the Influence of Computing on Mathematical Research and Education held at the University of Montana from August 14–17, 1973.

The invited addresses were directed primarily to the influence of the computer on mathematical research and the applications of mathematics and secondarily on what this means for the teaching of mathematics and the education of mathematicians. The contributed papers describe more specifically some experiments in developing courses in mathematics with computing and algorithmic orientations and a few reports on computer influenced research.

It is well known that the modern general-purpose computer makes possible computations hardly conceivable in the past. Our increased ability to compute and to handle data and information affects our daily lives and changes the very nature of the way in which mathematics is applied. Older classical mathematics is revived. Discrete mathematics is of growing importance. The need for new mathematical theories and new mathematical problems is generated by computing and by computer developments and applications. The computer intelligently used by, as yet, relatively few mathematicians has proved to be an important empirical tool for mathematical exploration and has provided the proof of theorems. All of this is reflected in the papers in this volume. Even where there has been little influence to date, the time seems near when in the hands of a new generation of mathematicians there will be a profound effect both in research and education of the constructive algorithmic approach to mathematics. This is certainly what all of those mathematicians who have become computer scientists tell us. Exploiting the pedagogical value of the computational and algorithm approach in the teaching of mathematics is still in a primitive and experimental state. It would seem unwise for mathematicians to close their minds to the possibilities. Fortunately, as this conference has shown, this is not likely to happen.

The sponsoring societies wish to thank all of the participants in the conference and particularly those who have contributed to this volume. The support of the National Science Foundation is gratefully acknowledged.

The topic of the conference was selected by the AMS-SIAM Committee on Applied Mathematics (Donald G. M. Anderson, Hirsh G. Cohen, Joaquin B. Diaz, Harold Grad, Stanislaw M. Ulam, and Richard S. Varga, chairman) and the officers of the American Mathematical Society and the Mathematical Association of America. The Organizing Committee included William S. Dorn (University of Denver), Stephen J. Garland (Dartmouth College), Thomas E. Hull (University of Toronto), Donald E. Knuth (Stanford University), and Joseph P. LaSalle (Brown University), chairman.

J. P. LaSalle
EDITOR

FEBRUARY, 1974

INVITED ADDRESSES

Proceedings of Symposia in Applied Mathematics
Volume 20
1974

THE INFLUENCE OF COMPUTING ON
RESEARCH IN NUMBER THEORY

BY

D. H. LEHMER

I am gratified to lead off this series of articles on the influence of computing on research by addressing you on the subject of this influence on the theory
of numbers. This is a subject that I have been observing since about 1908. My
father did many things to make me realize at an early age that mathematics, and
especially number theory, is an experimental science. If one examines the collect-
ed works of Euler, Gauss, Legendre, to name but three, one finds them shame-
lessly and laboriously computing examples of empirical discoveries. Often these
efforts led to the establishment of important theorems. Some of these discoveries remain to this day without logical links to Peano's axioms.

Exploring in discrete variable mathematics is generally simpler than in continuum mathematics. One can see the input and the resulting experimental output with absolute clarity. For the same reason a digital or discrete variable computer is a better aid to discovery than an analog machine. This advantage is due
to the enormous flexibility possessed by digital computers. Exploits such as moon
missions would be utterly impossible without discrete variable techniques, despite
the continuity of space. We should regard the digital computer system as an
instrument to assist the exploratory mind of the number theorist in investigating
the global and local properties of his universe, the natural numbers and their
algebraic extensions.

The role of this system in such an investigation can be of varying importance,
ranging from the production of a single counterexample, to the organization of
data to suggest ideas, through the search for patterns in data, to the ultimate role
of proving theorems on its own.

The role of the system can also be measured in terms of its expended effort,

AMS (MOS) subject classifications (1970). Primary 10–04, 10A35, 10A45, 68A20,
68A40.

that is in terms of machine time. Now-a-days this can be measured in dollars. There was a time when relatively large amounts of machine time, such as weekends or graveyard shifts, were given away for three nonaltruistic reasons:

(a) The machine would otherwise have been idle.

(b) The hardware suffers less from thermal stress and cathode poisoning when not turned off.

(c) Long runs with absolute checking at the end would reveal intermittent failures or confirm their absence.

In those days, certain computing establishments occasionally permitted long runs on problems in the theory of numbers. By way of illustrating such events I shall mention, with varying degrees of detail, examples dating from the very beginning to twenty years ago. Since 1953 there have been a number of developments, not all favorable, which have modified the influence of the computer. We shall not discuss this part of the history here.

The first such event took place in 1946. As you may recall, the Fourth of July fell on a Thursday that year. The ENIAC, the first electronic computer, was to have been shut down from Wednesday night till the following Monday morning. Instead, this chunk of 111 hours of machine time was made available to me and my wife in order to keep the ENIAC warm and active. The problem we decided to run was the following:

For each odd prime p there is a least positive integer $e = e(p)$ such that $2^e \equiv 1 \pmod{p}$, sometimes called the *order* or *exponent* of 2 modulo p. The problem proposed to the ENIAC was to find those primes p for which $e(p) \leqslant 2000$ until Monday morning at 8 o'clock. During the weekend the limit 2000 was reduced to 1000 and then to 300 in order to speed things up. By Monday, we had reached $p = 4538791$. This successful run had a number of consequences, even legal ones, which I shall not discuss. Instead, I shall say a few words about the program or algorithm the ENIAC used. For a given p a good deal is known in advance about $e(p)$. Thus e divides $p - 1$. Further examination of p may reveal that e divides $(p - 1)/2$ or $(p - 1)/3$, etc. Previously published tables of e were found to contain a great many errors, in spite of, or perhaps because of, this somewhat sophisticated approach to the problem.

In contrast, the ENIAC was instructed to take an "idiot" approach, based directly on the definition of e, namely, to compute

$$2^n \equiv \Gamma_n \pmod{p} \quad (n = 1, 2, 3, \cdots)$$

until the value 1 appears or until $n = 2001$, whichever happens first. Of course, the precedure was done recursively by the algorithm:

$$\Gamma_1 = 2, \quad \Gamma_{n+1} = \begin{cases} \Gamma_n + \Gamma_n & \text{if } \Gamma_n + \Gamma_n < p, \\ \Gamma_n + \Gamma_n - p & \text{otherwise.} \end{cases}$$

Only in the second case can Γ_{n+1} be equal to 1. Hence this delicate exponential question of finding $e(p)$ can be handled with only addition, subtraction, and discrimination at a time cost, practically independent of p, of about 2 seconds per prime. This is less time than it takes to copy down the value of p and in those days, this was sensational.

The next big job was run, on and off, over the next couple of years, on the same machine and on a related problem by George Reitwiesner, then of the Aberdeen Proving Ground where the ENIAC was finally installed. This was perhaps the first instance of the interrupted idle time modus operandi. The problem here was to find for each prime p the so-called Fermat quotient q_2 defined by

$$2^{p-1} - 1 \equiv pq_2 \pmod{p^2}.$$

Particular interest is attached to those p for which $q_2 = 0$, for only when this happens can there be solutions of Fermat's equation $X^p + Y^p = Z^p$ in integers X, Y, Z prime to p. Hand calculations by Beeger for $p < 16000$ showed no such examples except $p = 1093$ and 3511.

In 1947 the ENIAC was drastically altered from the most parallel computer ever built to a slower serial machine somewhat like the one bottleneck machines of today. This was done to shorten the time required to change from one problem to another. Thus it was possible to run "Slow Moses," as this program was called, whenever the ENIAC would have otherwise been standing idle for a time interval rather greater than that required to change a problem. The program was also of "idiot" type and this was a mistake, for it ran so slowly that after a couple of years p had reached only about 25000, as I recall. No other example of $q_2 = 0$ was found. This disappointment cannot fairly be laid at the feet of Slow Moses. We now know that the next example of $q_2 = 0$, if there is one, must correspond to a $p > 3,000,000,000$ [1].

Other electronic computers became available in the early 1950's and number theorists began to get some machine time. There was the SEAC on which K. Goldberg calculated the so-called Wilson's quotient W_p defined by

$$(p - 1)! + 1 \equiv pW_p \pmod{p^2}$$

for $p < 10000$, which also is connected with Fermat's Last Theorem. The only known examples of $W_p = 0$ are for $p = 5, 13$ and 563 [2].

There was an interesting but little known calculation by Lehman and Spohn done on the ORDVAC concerning the number of prime factors of integers. The question is: Among the integers $\leqslant x$, are those with an odd number of prime factors more popular then those with an even number of prime factors? According to a conjecture of Polya, which implies the Riemann Hypothesis, the "odds" are always at least as popular as the "evens" for $x \geqslant 2$. The ORDVAC verified this for all x up to several million, using a sophisticated bit pushing program. A decade later, Lehman, by a still more elaborate program, was able to find a value of x below which a majority of the integers are "evens," thus disproving Polya's conjecture [4].

An older conjecture of Kummer's (1846) concerns the cubic Gauss sum

$$x_p = \sum_{n=0}^{p-1} e^{2\pi i n^3/p} \qquad (p \equiv 1 \ (\mathrm{mod}\ 6)).$$

This number is a root of the cubic equation

$$x^3 - 3px - pL = 0$$

where L is uniquely determined by the representation

$$4p = L^2 + 27M^2, \qquad L \equiv 1 \ (\mathrm{mod}\ 3).$$

This cubic equation has three real roots, so the question arises: Which of these roots is x_p? Is x_p the least, middle, or largest root? Kummer put the 45 primes $p < 500$ into three categories accordingly, and found their cardinalities to be 7, 14 and 24. He therefore conjectured that the proportion of primes in the three categories tends to $1:2:3$. Using the IAS machine, Goldstein and von Neumann [3] extended Kummer's calculation to the first 611 primes and found the proportions more like $2:3:4$, namely 138, 201, 272. This shed considerable doubt on Kummer's conjecture. (The modern conjecture is that the 3 kinds of primes are asymptotically equidistributed.) This calculation was an interesting mixture of integer and real variables, and required about 20,000,000 multiplications.

With the establishment of the National Bureau of Standards' Institute for Numerical Analysis and the completion of its computer, the SWAC, in 1952, there came a small flood of research-oriented computations, some concerned with number theory. These have been described by Emma Lehmer in a paper entitled *Number theory on the* SWAC [6]. Sample topics are: Tests of Mersenne Primes, Riemann's Hypothesis, Fermat's Last Theorem, Kummer's Conjecture (in which the ratios $3:4:5$ were achieved), K-Sums, Restricted Permutations. These early efforts developed methods that led to a wider and more elaborate set of investigations on the same, or related, problems over the intervening two decades.

To discuss today's number-theoretic computations using modern computing systems, we find that only a few ideas need treatment.

With a given problem to investigate, one can write an ad hoc program. After doing this a few times with similar types of problems, one realizes that it is wise to prepare, once and for all, a family of subroutines for doing specific operations. Each member of this family can then be called upon if, and when, its peculiar operation is required. Without impairing seriously the overall efficiency of the resulting program, we work at a higher level of programming and make fewer mistakes. Some subroutines are written in machine independent language such as ALGOL or FORTRAN. While others are written in the machine's own assembly language and reflect its own characteristic specifications, such as the width of its arithmetic unit, the capacity of its registers and main memory or the available Boolean operations.

The standard format for information in such programs is the vector with integer components. The most frequently met difficulty is that of a component becoming too large to fit in its allotted space. The simplest example is that of an integer variable exceeding the capacity of the machine word. This happens very frequently in combinatorial number theory. Big machines today have 60 or 64 binary digits in a word but to do exact fixed-point integer arithmetic, we must limit ourselves to integers of 24 or 16 binary digits. To put it bluntly, modern machines are not designed for number theory. In self-defense, the arithmetician must resort to a family of subroutines that handle integers that occupy many words. These are the so-called multiprecision arithmetic subroutines.

The best method of representing a large integer I is via a signed absolute value format consisting of a vector of n one-word nonnegative integers together with an extra word (possibly contiguous) called the length of I containing $\pm n$ according as I is positive or negative. The length of I may vary during a calculation but it is always minimal; that is, the nth word of I is always non-zero. Only $I = 0$ has length zero. The address of I is taken as the actual address of its least significant word. The allocation of sufficient memory space for I is the user's only responsibility.

Besides the five arithmetic operations in the ring of integers, namely, addition, subtraction, multiplication, division with remainder, square root with remainder, there are custodial operations such as decimal to multiprecision input and its inverse for output, moving a copy of I from one address to another, etc., as well as Boolean operations and shift operations in multiprecision format. Finally, there are the number theory operations: GCD, Jacobi's symbol, Chinese remainder algorithm, power algorithm which finds the value of a^n (mod m), and several others.

Armed with this family, programming consists simply of a series of calls to subroutines as needed, interspersed with minor bookkeeping operations. We thus have a kind of language for number theory for numbers of any size.

The antithesis of multiprecision is "fractional precision," in which system several small integers are stored in one word. The compactification is an attempt to save both time and space. We are therefore talking about a rather sophisticated problem in a possibly less sophisticated language. The operations upon such vector words would not be multiplication or division but, more likely, logical operations like AND, OR, NAND, mask, shift, complement, and the like. Here we can use the full 60 or 64 bits in a word.

The extreme of fractional precision is the case in which each integer component has a single binary digit. That is, we are storing the values of a two-valued function $f(n)$ such as for example one of the numerical functions

$$f(n) = |\mu(n)| = \begin{cases} 1 & \text{if } n \text{ is square-free,} \\ 0 & \text{otherwise,} \end{cases}$$

$$f(n) = \begin{cases} 1 & \text{if } n \text{ is a product of an odd number of prime factors,} \\ 0 & \text{otherwise,} \end{cases}$$

$$f(n) = \begin{cases} 1 & \text{if } 2n + 1 \text{ is a prime,} \\ 0 & \text{otherwise,} \end{cases}$$

$$f(n) = \begin{cases} 1 & \text{if a theorem with parameter } n \text{ is true,} \\ 0 & \text{otherwise.} \end{cases}$$

In general

$$f(n) = \begin{cases} 1 & \text{if } n \in S, \\ 0 & \text{otherwise,} \end{cases}$$

where S is some given set of integers, finite or not. Here we are using, in the most compact format possible, a membership list for S. It answers without search the question: Does n belong to S? Although this question fails to excite professional set theorists, there are many practical uses for this device.

Such, then, are the data structures and space allotments of information for number theory problems. Besides space, one must consider time. For us, time is money. It goes without saying that programming in most subroutines should be polished to achieve the utmost in speed. Knowing the average execution time of the most frequently used subroutines allows us to guess whether a proposed program is feasible or not feasible. We say that an algorithm, or a subroutine, is of order $F(n)$, where n is one of its parameters, in case its execution time is less than a constant times $F(n)$. We can use F to make feasibility studies ignoring the constant multiplier. Thus, for example, if most of the time is to be spent in the GCD subroutine, we can predict with some assurance that doubling the number of digits in the numbers will only double the cost of our program. This is because the number of divisions required by Euclid's algorithm for $GCD(m, n)$ with $m > n$ is less than five times the number of decimal digits in n, a well-known theorem of Lamé. That is, GCD is of order $\log n$. Surprisingly many of the basic number theory algorithms are of this order, including the ubiquitous power algorithm, tests for primality, square root modulo p, etc.

Problems involving algorithms of higher order than $\log n$ often present difficulties to the unsponsored number theorist to whom free idle time on the big machine is not aviailable. One way out of this difficulty is to acquire one's own small, inexpensive computer. A number of such machines are now available in the price range that a mathematics department (until recent cutbacks) could well afford. Some of these computers are talented enough to serve as an aid to teaching a calculus or a numerical analysis course, using a BASIC compiler. They can also play Blackjack and Nim. These I do not recommend for the following reasons:

(a) They are poorly designed for number theory.

(b) They are too slow.

(c) They are too available to others.

A better move is to turn to a micproprogrammed computer which is still cheaper, much faster and, when properly programmed, is just what is needed. Also, it can be made troublesome for a calculus student. A system of this sort is being set up at Northern Illinois University by John Selfridge, to whom questions about details should be addressed.

Another more drastic approach to the same problem is to build one or more special purpose "off line" computers. This would be a small machine normally operating independently of the main computer but depending on the big machine for preparation of input and processing of output data, these interruptions requiring only a few seconds of expensive machine time. The off line machine can actually be faster than the big machine that we cannot afford to use anyway. Since the off line machine is fully dedicated and special, it is protected against intrusion by general users. Using integrated circuit technology, we can bring the design problem down to a level at which the number theorist can specialize to his own requirements.

Such an off line machine has been operating 24 hours per day for eight years without a budget at the University of California at Berkeley. A second machine is nearing completion. A third machine is in progress at the University of Illinois. These special machines, called sieves, are sufficiently general to deal with a wide class of problems such as solving $f(x, y) = 0$ where f is a polynomial with integer coefficients. In such a machine k different periodic patterns of pulses or nonpulses are being delivered to k teminals at each clock pulse time. If there is a pulse at each of the k terminals, the time is printed out. It is clear that such a simple machine can find the integer solutions to any problem whose solutions modulo p can be predicted in advance. Such problems abound in the theory of numbers.

In conclusion, let me return to the assigned title of this article: The influence of computing on research in number theory. I plan to give a fairly recent example of such influence in proving a simply stated theorem whose demonstration without machine aid appears to be humanly impractical. The theorem is

THEOREM. *Every set of 7 consecutive integers greater than 36 contains either a prime or a multiple of a prime greater than 41.*

That 7 cannot be replaced by 6 is shown by the factorizations

$$284 = 2^2 \cdot 71, \qquad 287 = 7 \cdot 41, \qquad 290 = 2 \cdot 5 \cdot 29,$$
$$285 = 3 \cdot 5 \cdot 19, \qquad 288 = 2^5 \cdot 3^2, \qquad 291 = 3 \cdot 97,$$
$$286 = 2 \cdot 11 \cdot 13, \qquad 289 = 17^2,$$

wherein we see 6 consecutive integers 285–290 having no prime factor > 41. Incidentally, it is true, but not obvious, that this situation will never happen again, beyond 285.

Number theory abounds with theorems, facts and conjectures about special infinite subsets of the natural numbers. Some of these have densities, like the square-free numbers with density $6/\pi^2$. Some do not, like the primes or the powers $(a^k, a > 1, k > 1)$. For these subsets, one can ask how the elements of the subset are distributed. In particular, one can ask whether, infinitely often, pairs of them are close together. In the case of the primes, we have the twin prime conjecture which says "yes." For the powers, we have the Catalan conjecture which says "no"; the accidents $9 - 8 = 1$ and $27 - 25 = 2$ will never happen again.

For our theorem we are considering the "round numbers" or, in Western's terminology, the A_n numbers, namely those numbers divisible by no prime exceeding the nth prime p_n. These numbers have been considered from time to time ever since Gauss (perhaps earlier) as they arise in the calculation of very accurate logarithms, in certain Diophantine equations including Fermat's equation, and in the distribution of consecutive power residues.

It is pretty clear that, for n fixed, the set of all A_n numbers has density zero. In fact, the number of A_n numbers $\leqslant x$ is asymptotic to

$$\frac{(\log x)^n}{n! \log 2 \log 3 \cdots \log p_n} = o(x).$$

We can then ask: Are there infinitely many pairs of A_n numbers differing by 1? In reply, Störmer (1897) [7] showed that, no matter how large n may be, there are only finitely many pairs of A_n numbers which are consecutive integers. Since $41 = p_{13}$, it is clear that we are concerned with A_{13} numbers and that the problem is that of showing that, beyond 36, there is no set of 7 A_{13} numbers that are consecutive integers.

It is also clear that if we could inspect a list of all instances of consecutive pairs of A_{13} numbers, we could easily ascertain the truth of our theorem. According to Störmer, this list is finite. All that remains is the discovery of this list and the proof that it is complete. This we leave to the machine.

To sketch this operation, let Q be the infinite set of all A_{13} numbers and let $Q' \subset Q$ be the finite subset of all square-free members of Q. The cardinality of Q' is of course $2^{13} = 8192$. Any pair of consecutive integers can be written $\frac{1}{2}(x - 1)$ and $\frac{1}{2}(x + 1)$ where x is an odd integer. For this pair to belong to our list, it is necessary that their product be an even A_{13} number that is

(1)
$$\frac{x - 1}{2} \frac{x + 1}{2} = 2D(y/2)^2$$

where $D \in Q'$ and $y \in Q$. But (1) can be written as

$$x^2 - 2Dy^2 = 1.$$

Conversely, every solution (x, y) of this Pell equation with $y \in Q$ gives an entry $\frac{1}{2}(x - 1)$, $\frac{1}{2}(x + 1)$ in our list. Thus we have only to solve the 8192 Pell equations (2) using the well-known method of the continued fraction development of the square root of $2D$, examining y for membership in Q. Unfortunately, for each $D \neq 2$ there are infinitely many solutions (x, y) of (2). Here the human comes to the rescue by knowing something of the divisibility properties of the y's. In fact, only the 21 smallest values of y that go with a given D can belong to Q. This fact finally reduces the calculation to a finite job. The last entry in our list of 869 pairs turns out to be $n, n + 1$ with $n = 63927525375 = 3^3 5^3 7^7 \cdot 23$ and $n + 1 = 2^{13} \cdot 11^4 \cdot 13 \cdot 41$. The preparation and proof of completeness of this list is quite beyond the capability of human beings, involving, as it does, hundreds of thousands of operations on multiprecise integers, each having hundreds of words [5].

The influence of the computer in the above example, though great, is relatively minor when compared to more elaborate programs that have been run. I refer to programs in which the course of the proof of a theorem is left to the discretion of the computer. The discussion of such techniques would take us far beyond our allotted space;

REFERENCES

1. J. D. Brillhart, J. Tonascia and P. Weinberger, *On the Fermat quotient*, Computers in Number Theory, Academic Press, New York, 1971, pp. 213–222.

2. K. Goldberg, *A table of Wilson quotients and the third Wilson prime*, J. London Math. Soc. 28 (1953), 252–256. MR 14, 1062.

3. H. H. Goldstein and J. von Neumann, *A numerical study of a conjecture of Kummer*, Math. Comp. 7 (1953), 133–134.

4. R. S. Lehman, *On Liouville's function*, Math. Comp. 14 (1960), 311–320. MR 22 #10955.

5. D. H. Lehmer, *On a problem of Störmer*, Illinois J. Math. 8 (1964), 57–79. MR 28 #2072.

6. E. Lehmer, *Number theory on the* SWAC, Proc. Sympos. Appl. Math., vol. 6, Numerical Analysis, McGraw-Hill, New York, 1956, pp. 103–108. MR 18, 74.

7. C. Störmer, *Quelques théorèmes sur l'equation de Pell et leurs applications*, Skrifter Videns.-Sels. I Mat.-Naturv. Kl. 1897, no. 2, 48 pp.

UNIVERSITY OF CALIFORNIA, BERKELEY

Proceedings of Symposia in Applied Mathematics
Volume 20
1974

THE INFLUENCE OF COMPUTERS ON ALGEBRA

BY

CHARLES C. SIMS

This article is supposed to describe the ways in which the branch of mathematics we call algebra has been influenced by the existence of large digital computers. To be complete, such a survey should include a description of the types of research problems which have been attacked using computers and illustrations of the sorts of results which have been obtained. However, during the last few years there have been several conferences at which reports have been given concerning results in algebra which have been obtained with the help of computers. The proceedings of four of these conferences are listed as references [2], [3], [5] and [9] at the end of this paper. The reader can find a great many examples of the use of computers in algebra by glancing through these proceedings. I shall therefore limit myself to a general discussion of computers in algebraic research and to some ideas concerning the way computers could be used in instruction in algebra.

A word should be said about what the term "algebra" is to mean for this article. There is a rapidly growing area of applied modern algebra which seeks to solve algebraic problems which arise in such fields as computer hardware design, computer software design and communications technology. My knowledge of this area is quite limited and I could not possibly improve on the excellent survey article on applied modern algebra by Garrett Birkhoff in [3]. I shall therefore not consider such topics as switching theory and coding theory here. My remarks will be confined to the use of computers to solve problems which arise naturally in the field of abstract algebra as discussed in such standard texts as [4] and [10].

There is no doubt that there has been significant use of the computer in algebraic research. The largest use of computers in algebra (as measured by a count of papers in the four references mentioned above) appears to have been

---

AMS (MOS) subject classifications (1970). Primary 00A25, 20-04, 20B05, 20K05; Secondary 02E10, 68A10.

made in finite group theory. One of the first algebraic algorithms to be pro-grammed for a digital computer was the Todd-Coxeter algorithm for coset enum-eration, a method for determining indices of subgroups of groups defined by generators and relations. The literature on the computer implementation of co-set enumeration goes back almost 20 years. The only existence proofs of several of the recently discovered finite simple groups rely heavily on computers. Out-side group theory, computers have been used to solve problems about finite fields, in commutative algebra and in algebraic topology. If combinatorial theory is included within algebra, then the considerable computer time spent on the problem of projective planes of order 10 should be mentioned.

However, after saying all this, I must conclude that the existence of the digital computer has not yet really changed the subject of abstract algebra, either in its content or in the way it is taught. The overwhelming majority of algebra-ists do not choose their research problems on the basis of whether or not a large computer or a particular sophisticated software package is available. The number of algebraists for whom computing plays a central role in their on-going research interests remains small.

More typical is the case of the algebraist who runs across a problem in his work which seems to require a computer solution. Usually there is no already-existing "canned program" to solve the problem and the algebraist must either write his own program or have it written by someone else. The algebraist probably does not know much about computing and the applications programmers available at the local computer center probably do not know much abstract algebra. As a result the program either does not get written or if it does it is often not as efficient as it could be. Considering the financial difficulties of may universities and their limited computing budgets, we all have an obligation to make effective use of computing.

There is a reason to believe that the use of computers in algebra will con-tinue to grow and that the programs written will be more and more sophisticated. An introduction to computing is becoming a standard part of an undergraduate mathematics major. This fact has two corollaries. First, there will be an increase in the number of students in advanced algebra courses who have a knowledge of computing. These students can help in the preparation of efficient programs to solve algebraic problems. Second, senior algebraists are being forced to learn enough about computing to teach undergraduate courses in which the computer is used. They will not, of course, become experts over night but the experience will certainly help them make better use of computers in their research.

There are many difficulties associated with writing programs to solve algebraic problems. Some, such as rapidly growing space requirements, are shared by a much wider class of problems. I would like to concentrate on two points which I think are particularly relevant to algebraic computing:

(1) The representation of algebraic objects.

(2) The novelty of algebraic algorithms.

Algebra has many ways of constructing new objects from old ones. Thus a group theorist might say: Let $H = A_6$, the alternating group of degree 6. Let $K = PGL(2, 7)$, the projective general linear group obtained as the quotient group of the group $GL(2, 7)$ of all nonsingular 2×2 matrices over the field of 7 elements modulo the group of nonzero scalar matrices. Finally, let $R = Z(H \times K)$, the integral group ring of the direct product of H and K. Elements of R are then formal integral linear combinations of ordered pairs, the first components of which are even permutations of the integers $1, 2, \cdots, 6$ and the second components of which are sets of six 2×2 matrices over the field of 7 elements. Clearly the problem of representing elements of R on a computer is not a trivial one.

Let us turn to the second point mentioned above. A great deal of computation was carried out long before the advent of the digital computer. Tables of transcendental functions were computed, astronomical tables were devised and lists of prime numbers were constructed. This work necessitated the development of relatively effective algorithms for solving certain types of algebraic and differential equations and for carrying out some number-theoretic constructions by hand.

In abstract algebra this tradition of hand computation is, to a large extent, lacking. As a result we often find ourselves having to work out for the first time algorithms of a quite elementary nature, elementary in the sense that they require no deep results to prove their correctness and could have been discovered 50 years ago had the motivation to find them been present.

This lack of interest in hand computation is evident in the way abstract algebra is taught. We expect every mathematics graduate student to know the fundamental theorem of finitely generated abelian groups. However, it is my experience from several years of giving Ph. D. oral exams that very few students can easily take a presentation of an abelian group such as

$$xy = yx, \qquad x^4 y^8 = x^8 y^4 = 1,$$

and express the group defined as a direct product of cyclic groups. Students are too often taught the fundamental theorem as an existence theorem and not as a computational algorithm.

The number of students who can solve the above problem is nevertheless large compared with the number who can solve the following problem: Let

$$G = \langle (1\ 2\ 3\ 4\ 5\ 6\ 7), (1\ 2\ 7\ 4\ 6\ 5\ 3) \rangle,$$

the group generated by the two given cyclic permutations. What is the order of G? There is a reasonably efficient algorithm to solve this problem which is conceptually no more difficult than the algorithm to determine the cyclic decomposition of a finitely presented abelian group.

The above examples are given to illustrate the point that it would be a good thing if algebra could be taught with more emphasis on constructive proofs and efficient algorithms for performing algebraic constructions in particular cases. This means giving plenty of "numerical" homework exercises. One difficulty with this approach is that beginning algebra students are unfamiliar with multiplying permutations or performing operations in finite fields. They are likely to make a great many errors in computation and become discouraged.

One solution to this difficulty is to have students write computer programs to solve their algebra homework. The success of this approach depends heavily on the computer language used, as anyone who has asked students in a linear algebra course to program Gauss elimination in FORTRAN can testify to. Relatively simple algebraic constructions require FORTRAN programs which are too complex for undergraduates to produce on a weekly basis.

There is a language which seems to be remarkably well suited both for instruction in what might be called computational abstract algebra and for a great deal of research computation in algebra. This is the language APL. APL is largely the creation of one man, Kenneth Iverson, and it takes its name from the title of the book [7] in which Iverson described the first version of his language.

This is not the place to try to teach APL. However, since I want to give some examples of how APL could be used in an algebra course, I shall give a very brief introduction to the language. Those readers who are not already familiar with APL and whose curiosity is sufficiently aroused to want to study the examples in detail should consult one of the growing number of books on APL, for instance [1], [6] or [8]. The reader is to be warned that the whole story is not being told in the discussion which follows. In particular many of the APL operators mentioned have more general definitions than those stated here.

APL is implemented in a time-sharing environment. Let us assume that we are seated at an APL terminal and that we have just logged on. The carriage is automatically positioned so that our input lines are indented 6 spaces while

responses by the computer begin at the left margin. In the sample dialogues
given here the left margin and the 6 space indentation will be represented like this:

FIGURE 1

Consider the following short dialogue:

```
      I←2
      J←3
      I+J
5
```

FIGURE 2

Here we see that the left-pointing arrow ← is used to assign values to variables.
Also, in the normal mode of operation, APL expressions are evaluated immediately
and if no assignment of the result is indicated, then the result is printed at the
terminal. We continue.

```
      10+4-3
11
      10-4+3
3
      (10-4)+3
9
```

FIGURE 3

The second result may seem incorrect. This is because APL differs from traditional
mathematical notation in the order in which operations are performed. APL has
many operators besides the usual arithmetic ones. It was found impossible to
devise reasonable precedence rules to govern the order of applying the operators.
Instead a simple right-to-left rule was adopted. In the absence of parentheses,
operations are performed from right to left. Thus in evaluating the expression

$10 - 4 + 3$, the computer first adds 4 and 3 and then subtracts the result from 10.

¯10	10-20
¯10	-10
¯7	-4+3
¯1	¯4+3
1	×10
¯1	×¯10
0	×0
3	|3
3	|¯3
1	2=2
0	2<1
3	⌊3.5
¯4	⌊¯3.5
3	3⌊4
¯4	¯3⌊¯4

FIGURE 4

Notice the difference between the symbol used in the representation of a negative constant and the symbol used to denote the operations of subtraction and negation. This is necessary because of the right-to-left rule for evaluation. In order to keep the number of symbols to a minimum, one symbol often is used to denote two functions, one a monadic function, with one argument, and the other a dyadic function, with two arguments. Thus the multiplication symbol × is used to denote the monadic signum function which is 1, − 1 or 0 according as its argument is positive, negative or 0. The symbol | used monadically denotes absolute value. Logical values of true and false are represented by 1 and 0, respectively. The symbol ⌊ represents the monadic greatest integer function and the dyadic minimum.

One of the great strengths of APL is the ease with which arrays can be handled.

```
            |V←1 2 3 5 8 13
            |V
  1   2     |3   5   8   13
            |V[4]
  5         |
            |V[2 4 6]
  2   5     |13
            |Vι2 5 13
  2   4     |6
            |ι5
  1   2     |3   4   5
            |M←2 3ρV
            |M
      1     | 2    3
      5     | 8   13
            |M[2;1]
  5         |
            |M[1 2;2 3]
      2     | 3
      8     |13
            |ρV
  6         |
            |ρM
  2   3     |
            |,M
  1   2     |3   5   8   13
            |W←1 3 5
            |V,W
  1   2     |3   5   8   13   1   3   5
            |φV
 13   8     |5   3   2   1
            |1↓V
  2   3     |5   8   13
            |Vο.×W
      1     | 3    5
      2     | 6   10
      3     | 9   15
      5     |15   25
      8     |24   40
     13     |39   65
```

FIGURE 5

The components of a constant vector need not be enclosed in parentheses nor separated by commas. Indices are enclosed in square brackets. Indices of one array may be arrays themselves. The dyadic function $ι$ gives the index in the first argument of each entry in the second argument. Monadically $ι$ produces the vector of integers from 1 up to the argument. The dyadic function $ρ$ shapes a vector into a matrix or an array of higher rank. Used monadically $ρ$ returns the length of a vector or the shape of a matrix, that is, the number of rows and columns of the matrix. The monadic , forms a vector out of the entries of a matrix while the dyadic , catenates vectors to form longer vectors. The monadic function $φ$ reverses the order of the components of a vector. The function $↓$ permits the dropping of one or more components from a vector. The compound symbol $\circ.\times$ denotes the operation of forming all possible products of the entries of the first argument with entries of the second.

It is easy to write and store programs or subroutines in APL. The usual term for these is "function" but "procedure" would be a better name. Here is a simple example:

```
             ∇Z←X SUM Y
    [1]      Z←X+Y
    [2]      ∇
             2 SUM 3
    5
             10 SUM 4 SUM ¯3
    11
             10 SUM -4 SUM 3
    3
             10 SUM ¯4 SUM 3
    9
```

FIGURE 6

The first ∇ terminates the execution mode and begins the definition mode in which statements entered are not executed immediately but saved for future use. The rest of the first line indicates that we are defining a procedure SUM with two arguments X and Y which returns an explicit result Z. The system prompts with line numbers and we enter the statements defining the procedure. Entering another ∇ returns us to the execution mode. Note that the right-to-left rule applies to SUM also.

The ability to use one array as an index in another array makes computing with permutations quite easy.

```
             X←2 3 5 1 4
             Y←3 1 5 2 4
             X[Y]
    5   2    4   3   1
             Y[X]
    1   5    4   3   2
             ⍋X
    4   1    2   5   3
             X[⍋X]
    1   2    3   4   5
             (⍋X)[X]
    1   2    3   4   5
```

FIGURE 7

Because $X[Y]$ denotes the result of applying Y first and then X, it is convenient to think of permutations as acting on the left rather than on the right as is more traditional in the study of permutation groups. When applied to a permutation the operator $\oint$ gives the inverse permutation.

The many primitive functions in APL reduce the need for looping.

```
            A←1  0  0  1  1  0
            B←0  1  1  1  0  1
            V
  1    2    3    5    8    13
            A/V
  1    5    8
            B/V
  2    3    5    13
            +/V
  32
            +/A
  3
            v/A
  1
            ^/A
  0
```

FIGURE 8

If A is a logical vector, a vector of 0's and 1's, and if V is any vector of the same length, then A/V is the vector of components in V corresponding to 1's in A. The operation $+/$ computes the sum of the components in a vector. Analogously, other operations can be defined by replacing the "+" by any other primitive dyadic scalar function, for example the logical *or* function $\lor$ or the logical *and* function $\land$.

I would like now to define two procedures which will be used in the examples which follow. Here is the first:

```
        ∇C←A PROD B;I
[1]     C←(ρA)ρ0
[2]     I←0
[3]     →((ρA)[1]<I←I+1)/0
[4]     C[I;]←A[I;B[I;]]
[5]     →3
[6]     ∇
        A←2 3ρ1 3 2 2 3 1
        A
  1   3   2
  2   3   1
        B←2 3ρ2 1 3 3 1 2
        B
  2   1   3
  3   1   2
        C←A PROD B
        C
  3   1   2
  1   2   3
```

FIGURE 9

A set of permutations can be represented conveniently by a matrix whose rows give the permutations in the set. Given two such matrices of permutations A and B of the same shape, the procedure PROD finds the matrix C whose Ith row is the product of the Ith row of A with the Ith row of B. There are two branch statements in the definition of PROD. The fifth statement is an unconditional branch to statement 3. Statement 3 is a conditional branch of the form

$$\rightarrow (\text{condition})/\text{number}$$

which denotes a branch to the indicated statement if the condition is true. In statement 3 the value of I is increased by 1 and if I is greater than the number of rows of A, then a branch to statement 0 is made. Since statement 0 does not exist, this terminates the procedure. Placing I in the first line of the definition separated from the rest by a semicolon indicates that I is a local variable, one having meaning only within PROD, and not a global variable.

The second procedure generalizes the dyadic function ι.

```
          ∇I←A IOTA X
    [1]   I←1+(ρA)⊤¯1+(,A)ιX
    [2]   ∇
          D←2 3ρ1 2 3 4
          D
      1   2   3·
      4   1   2
          D IOTA 4
    2   1
          D IOTA 2
    1   2
```

FIGURE 10

When A is a matrix and X is a scalar the procedure IOTA returns the row and column of the first occurrence of X as an entry in A. IOTA is based on the encode function $\top$, whose definition I shall omit.

We come now to the first of the two examples I wish to give of the use of APL in teaching algebra. I want to sketch a proof of the existence part of the fundamental theorem of finitely generated abelian groups which at the same time demonstrates the validity of an APL procedure for actually computing cyclic decompositions of finitely generated abelian groups.

Recall the integer row operations on an integer matrix C.

1. Multiply a row of C by ± 1.

2. Interchange two rows of C.

3. Add an integer multiple of row i to row j, $i \neq j$.

We can define one-line APL procedures for carrying out these operations. For example:

```
        ∇M ROWOP1 I
[1]     C[I;]←M×C[I;]
[2]     ∇
        C←3 4ρι12
        C
    1    2    3    4
    5    6    7    8
    9   10   11   12
       ¯1 ROWOP1 2
        C
    1    2    3    4
   ¯5   ¯6   ¯7   ¯8
    9   10   11   12
```

FIGURE 11

Here the first argument of ROWOP1 gives the multiplier, which must be ± 1, and the second argument gives the number of the row to be multiplied. The other procedures are defined as follows:

```
        ∇ROWOP2 V
[1]     C[V;]←C[⌽V;]
[2]     ∇
        C
    1    2    3    4
   ¯5   ¯6   ¯7   ¯8
    9   10   11   12
        ROWOP2 1 3
        C
    9   10   11   12
   ¯5   ¯6   ¯7   ¯8
    1    2    3    4
```

FIGURE 12

The single argument of ROWOP2 in Figure 12 is a vector of length 2 which lists the rows to be interchanged. The definition of ROWOP3 in Figure 13 allows several multiples of a fixed row to be added to various other rows. In the first application of ROWOP3 the third row is added to the second. In the second application 9 times the third row is subtracted from the first and 4 times the third row is added to the second.

```
        ∇M ROWOP3 V
[1]     C[1↓V;]←C[1↓V;]+(,M)∘.×C[V[1];]
[2]     ∇
        C
   ¯9   10   11   12
   ¯5   ¯6   ¯7   ¯8
    1    2    3    4
        1 ROWOP3 3 2
        C
   ¯9   10   11   12
   ¯4   ¯4   ¯4   ¯4
    1    2    3    4
        ¯9 4 ROWOP3 3 1 2
        C
    0   ¯8  ¯16  ¯24
    0    4    8   12
    1    2    3    4
```

FIGURE 13

For purpose of comparison here is a FORTRAN subroutine for performing the same operations as ROWOP3.

```
        SUBROUTINE  ROWOP3(C, K, L, M, LM, I, V)
        INTEGER  C, V
        DIMENSION  C(K, L), M(LM), V(LM)
        IF ((LM. LE. 0)  OR  (L. LE. 0))  RETURN
        DO 20 J = 1, LM
              N = V(J)
              DO 10 JJ = 1, L
                    C(N, JJ) = C(N, JJ) + M(J) * C(I, JJ)
10                  CONTINUE
20            CONTINUE
        RETURN
        END
```

The integer column operations and APL procedures for applying them are similarly defined.

```
        ∇M COLOP1 I
[1]     C[;I]←M×C[;I]
[2]     ∇

        ∇COLOP2 V
[1]     C[;V]←C[;⌽V]
[2]     ∇

        ∇M COLOP3 V
[1]     C[;1↓V]←C[;1↓V]+C[;V[1]]∘.×,M
[2]     ∇
```

FIGURE 14

Two integer matrices of the same shape shall be said to be equivalent if one can be obtained from the other by a sequence of row and column operations.

Let us suppose that the existence part of the fundamental theorem of finitely generated abelian groups has been shown to be equivalent with the following:

THEOREM. *Any integer matrix* A *is equivalent to a matrix* C *of the form*

$$\begin{bmatrix} d_1 & & & & 0 \\ & \ddots & & & \\ & & d_r & 0 & \\ & & & \ddots & \\ 0 & & & & \ddots \end{bmatrix}$$

where each d_i *is positive and* d_i *divides* d_{i+1}, $1 \leqslant i < r$.

Before proving the theorem, we list an APL procedure for actually finding the matrix C. The proof of the theorem and of the correctness of the procedure will proceed together. The parts of the proof will be labeled by the statement numbers of the statements in the procedure to which they correspond.

```
      ∇ C←FUNDTHM A;M;N;B;X;I;K;L;Y;J
  [1]   C←A
  [2]   →(∧/,C=0)/0
  [3]   M←(ρC)[1]
  [4]   N←(ρC)[2]
  [5]   X←⌊/(,B≠0)/,B←|C
  [6]   I←B IOTA X
  [7]   ROWOP2 1,I[1]
  [8]   COLOP2 1,I[2]
  [9]   (×C[1;1]) ROWOP1 1
 [10]   (-⌊C[1;K]÷X) COLOP3 1,K←1+ιN
 [11]   →(∨/C[1;K]≠0)/5
 [12]   (-⌊C[L;1]÷X) ROWOP3 1,L←1+ιM
 [13]   →(∨/C[L;1]≠0)/5
 [14]   →(∧/,Y←0=X|C)/19
 [15]   J←Y IOTA 0
 [16]   1 ROWOP3 1,J[1]
 [17]   (-⌊C[J[1];J[2]]÷X) COLOP3 1,J[2]
 [18]   →5
 [19]   C[L;K]←FUNDTHM C[L;K]
      ∇
```

FIGURE 15

PROOF. (1) Let C be A initially. We shall apply row and column operations to C to bring it into the proper form. The proof will proceed by induction on the number of rows of C. Note that correspondingly the procedure is recursive due to statement 19.

(2) If C is the zero matrix, then C has the required form and we are done. This statement also terminates the recursion at the point that C has no rows.

(3) Let C have m rows.

(4) Let C have n columns. We now assert that by applying row and column operations to C we may bring it to the following form:

$$\left[\begin{array}{c|c} x & 0 \\ \hline 0 & C' \end{array}\right]$$

where x divides every element of C'. In the procedure this corresponds to the assertion that control will flow to statement 19 and when it does, then C has the indicated form. These assertions are proved by induction on the smallest of the absolute values of the nonzero entries of C.

(5) Let x be the smallest of the absolute values of the nonzero entries of C.

(6) Choose i_1 and i_2 so that $x = |C_{i_1 i_2}|$.

(7) Interchange rows 1 and i_1.

(8) Interchange columns 1 and i_2. Now $x = |C_{11}|$.

(9) Multiply the first row of C by -1, if necessary, so that $x = C_{11}$.

(10) For $2 \leqslant k \leqslant n$ write $C_{1k} = e_k x + f_k$, where e_k and f_k are integers and $0 \leqslant f_k < x$. Subtract e_k times the first column of C from the kth column. The first row of C is now

$$x f_1 f_2 \cdots f_n.$$

(11) If some $f_k \neq 0$, then we can apply induction on x since we now have a nonzero entry in C with absolute value less than x. Applying induction means going back to step 5. This will not happen if $x = 1$, for then all the f_k will be 0.

(12) For $2 \leqslant l \leqslant m$ write $C_{l1} = g_l x + h_l$, where g_l and h_l are integers and $0 \leqslant h_l < x$. Subtract g_l times the first row of C from the lth row. (The effect is simply to replace C_{l1} by h_l.)

(13) If some $h_l \neq 0$, then we can apply induction on x again. This will not happen if $x = 1$.

(14) The form of C is now

$$\left[\begin{array}{c|c} x & 0 \\ \hline 0 & C' \end{array}\right].$$

If x divides every entry of C, and hence of C', then we have proved the assertions in step 4 and we may go to statement 19.

(15) Choose j_1 and j_2 so that x does not divide $C_{j_1 j_2}$.

(16) Add the first row of C to the j_1th row. This makes $C_{j_1 1} = x$.

(17) Write $C_{j_1 j_2} = ux + v$, where u and v are integers and $0 \leq v < x$. Here $v \neq 0$. Subtract u times the first column of C from the j_2th column.

(18) We now have $C_{j_1 j_2} = v$ and we can apply induction on x.

(19) We now have C in the form asserted in step 4. By induction on the number of rows of C we can apply row operations and column operations to C' so that C has the form

$$\left[\begin{array}{ccccc} x & & & & 0 \\ & d_2 & & & \\ & & \ddots & & \\ & & & d_r & \\ & & & & 0 \\ 0 & & & & & \ddots \end{array}\right]$$

where the d_i are positive and d_i divides d_{i+1}, $2 \leq i < r$. Since the d_i are integral linear combinations of the entries of C' and x divides each entry of C', we know that x divides d_2 and so we have proved the theorem.

We can use FUNDTHM to solve the problem given above.

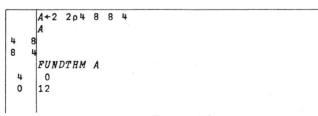

```
      |A←2 2ρ4 8 8 4
      |A
  4  8|
  8  4|
      |FUNDTHM A
  4  | 0
  0  |12
```

FIGURE 16

Thus the group defined is isomorphic to $Z_4 \times Z_{12}$.

When actually using FUNDTHM at a terminal, one must be careful because of the possibility of overflow. However, for small examples, say 5×5 matrices with entries no larger than 10 in absolute value, this will not be a problem.

The final example I want to discuss is an APL procedure ORDER which can be used to solve the second problem mentioned earlier, determining the order of the group generated by a set of permutations of relatively small degree.

ORDER uses three procedures in addition to **PROD** defined earlier. The first two are quite straightforward.

```
        ∇A←SORTOUT B;T
[1]     T←(⌈/B)ρ0
[2]     T[B]←1
[3]     A←T/ιρT
[4]     ∇
        SORTOUT 5 4 3 2 1 2 3
1    2  3    4    5

        ∇A←INV B;I
[1]     A←(ρB)ρI←0
[2]     →((ρB)[1]<I←I+1)/0
[3]     A[I;]←▲B[I;]
[4]     →2
[5]     ∇
        B←2 4ρ2 3 4 1 1 4 2 3
        B
    2  3  4    1
    1  4  2    3
        INV B
    4  1  2    3
    1  3  4    2
```

FIGURE 17

The procedure SORTOUT takes a vector of positive integers, deletes duplications among the components and arranges the resulting components in increasing order. The procedure INV takes a matrix B whose rows are permutations and returns the matrix A whose Ith row is the inverse of the Ith row of B.

The third procedure used by ORDER is a little more complicated.

```
      ∇ U←I REPS G;N;M;T;V
[1]     U←(1,N)ριN←(ρG)[2]
[2]     M←1
[3]     →(M=ρT←SORTOUT(V←U,[1]((M×(ρG)[1]),N)ρ,G[;U])[;I])/0
[4]     M←(ρU←V[V[;I]ιT;])[1]
[5]     →3
      ∇
        G←1 4ρ2 3 4 1
        G
    2  3  4    1
        1 REPS G
    1  2  3    4
    2  3  4    1
    3  4  1    2
    4  1  2    3
```

FIGURE 18

Let G be a matrix whose rows are permutations of $1, 2, \cdots, N$ and let I be an integer between 1 and N. The statement

$$U \leftarrow I \; REPS \; G$$

creates a matrix whose rows are permutations in the group G generated by the rows of G and these permutations are a set of left coset representatives for H in G, where H is the stabilizer of I in G. Said another way, the entries of the column $U[\;;I]$ are distinct and every point in the orbit of G containing I occurs as an entry in $U[\;;I]$. The algorithm used to construct U is rather crude but is sufficient for our purposes. Its steps are as follows.

(1) Initially let U be the matrix with one row consisting of the identity permutation.

(2) M will denote the number of rows of U. At first M is 1.

(3) Let V be the matrix of permutations consisting of the rows of U and all permutations obtained as a product of a row of G with a row of U. Let T be the set of (distinct) images of I under the rows of V. If T has M elements, then we are done.

(4) If T has more than M elements, then for each element of T choose a row of V taking I to that element and let U be the matrix of these permutations. Replace M by its new value.

(5) Go back to step 3.

We can now list the procedure ORDER.

```
      ∇  O←ORDER  G;N;I;H;U;J;W;V;X
  [1]     V←((N-1),N,N←I+(ρG)[2])ρ0
  [2]   ST1:→(0=I←I-1)/ST4
  [3]     H←(∧/G[;ιI-1]=((ρG)[1],I-1)ριI-1)≠G
  [4]     V[I;U[;I];]←INV U←I REPS H
  [5]     W←(((ρH)[1]×(ρU)[1]),N)ρ,H[;U]
  [6]     J←I-1
  [7]   ST2:→(N=J←J+1)/ST1
  [8]     →((ρW)[1]<X←V[J;W[;J];1]ι0)/ST3
  [9]     G←G,[1] W[X;]
  [10]    I←J+1
  [11]    →ST1
  [12]  ST3:W←V[J;W[;J];] PROD W
  [13]    →ST2
  [14]  ST4:O←+×/+/V[;;1]≠0
      ∇
        G←2 7ρ2 3 4 5 6 7 1 2 7 1 6 3 5 4
        G
    2  3  4  5  6  7  1
    2  7  1  6  3  5  4
        ORDER G
  168
```

FIGURE 19

Thus the group $\langle(1\ 2\ 3\ 4\ 5\ 6\ 7), (1\ 2\ 7\ 4\ 6\ 5\ 3)\rangle$ has order 168.

The procedure ORDER returns the order of the group generated by the rows of its argument G. What follows is a rough sketch of the algorithm used.

Let G be the group generated by the rows of G, which we assume are permutations of $1, 2, \cdots, n$. For $1 \leqslant i \leqslant n$ let $G^{(i)}$ be the group generated by those rows of G fixing $1, 2, \cdots, i-1$. For $1 \leqslant i < n$ let $U^{(i)}$ be a set of left cosets representatives for $G^{(i+1)}$ in $G^{(i)}$. If $G^{(i)}$ happens to be the stabilizer of $1, \cdots, i-1$ in G, then

$$G = U^{(1)}U^{(2)} \cdots U^{(n-1)} \quad \text{and} \quad |G| = \prod_{i=1}^{n-1} |U^{(i)}|.$$

It is a fact that each $G^{(i)}$ is the stabilizer of $1, \cdots, i-1$ in G provided that for $1 \leqslant i < n$, for u a row of $U^{(1)}$ and g a row of G fixing $1, \cdots, i-1$ the product gu is in the set

$$U^{(i)}U^{(i+1)} \cdots U^{(n-1)}.$$

This condition is tested. If it fails, then a new row is added to G and the process repeated until the condition does hold.

REFERENCES

1. APL\360 *user's manual*, IBM, White Plains, N. Y., 1970 (GH20-0683).

2. A. O. L. Atkins and B. J. Birch (Editors), *Computers in number theory*, Academic Press, London, 1971.

3. G. Birkhoff and M. Hall (Editors), *Computers in algebra and number theory*, American Math. Soc., Providence, R. I., 1971.

4. G. Birkhoff and S. Mac Lane, *A survey of modern algebra*, Macmillan, New York, 1965. MR 31 #2250.

5. R. F. Churchhouse and J. C. Herz (Editors), *Computers in mathematical research*, North-Holland, Amsterdam, 1968. MR 38 #1972.

6. L. Gilman and A. J. Rose, APL\360; *An interactive approach*, Wiley, New York, 1970.

7. K. E. Iverson, *A programming language*, Wiley, New York, 1962. MR 26 #913.

8. H. Katzan, APL *user's guide*, Van Nostrand-Reinhold, New York, 1971.

9. J. Leech (Editor), *Computational problems in abstract algebra*, Pergamon, Oxford, 1970. MR 40 #5374.

10. B. L. van der Waerden, *Moderne algebra*. Vols. I, II, Springer, Berlin, 1930, 1931; English transl., Ungar, New York, 1949, 1950. MR 10, 587.

RUTGERS UNIVERSITY

Proceedings of Symposia in Applied Mathematics
Volume 20
1974

COMPUTATIONAL PROBABILITY AND STATISTICS [1]

BY

ULF GRENANDER [2]

In 1968 John Tukey and Frederick Mosteller stated about the influence of the computer:

> Ideally we should use the computer the way one uses paper and pencil: in short spurts, each use teaching us—numerically, algebraically, or graphically—a bit of what the next use should be. As we develop the easy back-and-forth interaction between man and computer today being promised by time-sharing systems and remote consoles, we shall move much closer to this ideal.

Today, in 1973, we are indeed closer to this ideal. Although the mode of operation described by Tukey and Mosteller has not yet been widely adopted, it has become clear to a widening circle of practitioners of mathematical statistics that we can and must use the new possibilities of interactive computing.

At Brown University we initiated a project in 1967 called "Computational probability." [3] Its purpose was to explore the consequences for this area of applied mathematics of technological advances, not just the increased speed and storage capacity of the computer but also from time-sharing systems, new programming tools, and graphical displays.

During the academic year 1968–69 a graduate course was offered in computational probability to test some of the teaching material that had been developed.

AMS (MOS) subject classifications (1970). Primary 62–02.

[1] Figures and portions of text are reprinted with permission from *SIAM Review* (1973). Copyright 1973 by Society for Industrial and Applied Mathematics.

[2] This paper was delivered by Professor Richard A. Vitale, Division of Applied Mathematics, Brown University.

[3] The Computational Probability Project has been described in a series of reports that can be obtained from the Division of Applied Mathematics, Brown University, Providence, Rhode Island 02912.

At the same time we let the students investigate a number of research problems, some of which led to M. A. or Ph. D. theses. We encouraged an experimental approach, using the computer as the physicist uses his laboratory.

This course has afterwards been modified and repeated on different levels. The main body of this paper will describe in some detail several of the problems studied.

Let us start by describing how we tried to implement the program inasfar as teaching was concerned. Examples were presented, either in lectures together with the computational results or, more often, handed out to the students with some hints about possible ways of approaching the problem and asking them to do what they could to solve it. The examples presented were chosen because they would give the student a feeling for the degree of mathematical complexity that often arises in real-life applications as contrasted with the well-behaved textbook examples tailored to fit existing theory. A certain amount of time has to be spent presenting the subject matter background, but this is unavoidable if one wishes to describe and criticize the crucial relation between model and reality. Otherwise one could easily gloss over some of the messy details in the problem and smooth out realistic features of it. All models represent idealizations to some degree and a difficult phase in the model-building process is the choice of what features should be accounted for in the model and what one feels can be left out safely. How to strike this balance should be discussed in class.

Among examples of this type let us briefly mention a few. Many of the fundamental limit theorems were studied both by simulations and by direct computing. Simulation is usually easier but if has the drawback that the accuracy is low for reasonable computing cost. Direct computation is harder, involving some numerical analysis of the algorithms involved.

Particular attention was paid to situations where the limit theorems did not apply, or at least the convergence would be so slow as to make them practically meaningless. The laws of large numbers, the ergodic theorem for stationary stochastic processes, and the central limit theorems were among the ones studied in this way.

We tried to link up the limit theorem with the behavior of the stochastic processes associated with the partial sums. It was noticed early in the project that the didactic value increased dramatically when the results were displayed graphically. In the beginning this was done by simple plots produced by the terminal typewriter; later we experimented with plotters and cathode-ray tubes. More about this later.

The mathematical experiments and the phenomena displayed were later put into the right perspective by discussing the proof of the theorem in question. In this way the student got into the habit of correlating the results of the mathematical experiments with theory and using the computer as a tool to be combined with deduction, not to replace it.

In these instances the underlying theory was known but in many other cases this was not so: either the theory had not yet been developed and we tried to do it, or the model was so complicated that a purely deductive approach seemed hopeless.

A number of sequential situations were treated: search problems, sequential design problems, and a mathematical management game. The rules of the game were quite lifelike, describing a reinsurance activity, making the game so complex that any full game-theoretic study was out of the question. Instead, some of the graduate students tried to construct algorithms of the heuristic programming type which would imitate and possibly surpass human judgment and intuition. Later the program was made to play against several groups of students in the classroom, and we even matched different programs against each other. Time does not allow a detailed description of how the algorithms were derived; suffice it to say that they were essentially optimization techniques based on partial and possibly incorrect information, starting from a Bayesian approach.

The session in class where interactive programs alternated with human decision-making and analytical derivations led to some lively discussions, and it was clear that the students became personally involved and activated in a way that is otherwise rare, unfortunately.

This was all on the graduate level in the beginning of the project, and it is not so easy to draw a clear line between graduate instruction of the above type and research.

Turning to the research side, the approach was very much the same but the problems dealt with were now more extensive and led often to theoretical developments. Let us mention a few cases here.

Suppose that $x_t = m_t + y_t$ is a discrete-time stochastic process, $t = 1, 2,$ $\cdots$, where y_t is a wide-sense stationary process with zero mean value and co-variance matrix R and where, for each t, m_t is linear in s (constant) unknowns $\gamma_1, \cdots , \gamma_s$, so that, for $s < n$,

$$
(1) \qquad m = \Phi\gamma, \quad m = \begin{pmatrix} m_1 \\ \cdot \\ \cdot \\ \cdot \\ m_n \end{pmatrix}, \quad \gamma = \begin{pmatrix} \gamma_1 \\ \cdot \\ \cdot \\ \cdot \\ \gamma_s \end{pmatrix},
$$

$$\Phi = \text{a known } n \times s \text{ constant matrix of rank } s.$$

An estimate $d = (d_1, \cdots, d_s)^*$ (where $*$ means transpose) of γ which is linear in the observed vector $x = (x_1, \cdots, x_n)^*$ can be formed in at least two important ways. The least-squares estimate requires the minimization of $(x - \Phi d)^*(x - \Phi d)$ over all d. Under the assumption that $\Phi^*\Phi$ is nonsingular, this estimate is given by $d_{LS} = (\Phi^*\Phi)^{-1}\Phi^*x$. Since $E\ d_{LS} = (\Phi^*\Phi)^{-1}\Phi^*\Phi\gamma = \gamma$, there is no bias. A second and more fundamental approach is to seek that unbiased linear estimate which has minimum variance. If R is nonsingular, such a best linear unbiased (BLU) estimate is given by $d_{BLU} = (\Phi^*R^{-1}\Phi)^{-1}\Phi^*R^{-1}x$. One would like to find an estimate $\hat{d}$ which is *asymptotically efficient* in the sense that $\text{Var}(d_{BLU})/\text{Var}(\hat{d})$ tends to 1 as n tends to infinity.

The results which follow concern the special case $x_t = m + y_t$, where m is a fixed constant and y_t has a continuous spectral density. This special case illustrates well the difficulty that we meet and the way out of it. We have $s = 1$, $\gamma = m$ and $\Phi = (1, \cdots, 1)^* = e, n \times 1$. Linear estimates of m in this case are of the form $\Sigma_1^n c_\nu x_\nu$. The least-squares estimate is here of course the same as the straight arithmetic mean

$$(2) \qquad\qquad m_{LS} = \frac{1}{n}\sum_1^n x.$$

Let us also recall that the best unbiased linear estimate m_{BLU} reduces to

$$(3) \qquad\qquad m_{BLU} = c^*x,$$

where

$$(4) \qquad\qquad c = (e^*R^{-1}e)^{-1}R^{-1}e.$$

It is not difficult to see that under the stated conditions the BLU estimate is unique with

$$(5) \qquad\qquad C_{n+1-\nu} = C_\nu, \qquad \nu = 1, 2, \cdots, n.$$

The usefulness of the estimate m_{BLU} which is the best one theoretically can be questioned on the following grounds. First, the expression for m_{BLU} involves the inversion of the covariance matrix R. Now, if the sample size is large, say hundreds of observations, this inversion takes some effort. Except for some situations when R is very simple and the inverse can be obtained in closed form, it has to be done numerically. With the computing power that has been made available this can certainly be done, but even today it can be questioned whether it does not represent a waste of machine time. Another objection to using m_{BLU}, more intrinsic than the first one, is that R is seldom given a priori and is therefore not directly available to us unless we are willing and able to estimate it from data.

It has been shown, under certain conditions, that m_{LS}, which can be directly calculated, is asymptotically efficient. Later on this result was extended to general linear regression problems in time-series analysis and by now we have a fairly good understanding of these problems. It now constitutes an elegant and useful chapter of time-series analysis.

In the beginning of the development of this theory, in the 1950's, little computational work was done to find out what rate of convergence to the optimum we might expect. Today we can do it with comparatively little machine time. It is instructive to see, for instance, how the variance decreases and the efficiency tends to 1. To really make use of this information we have displayed it graphically by means of a Calcomp plotter. A few of these graphs are given in Figures 1–6.

Some experiments will be illuminating. We choose a spectral density f, calculate the optimal coefficients C_ν and graph them against ν as well as the efficiency of m_{LS} against n. Note that here there is no point in employing simulation, since it is computationally more efficient to do the needed matrix inversion directly.

For the spectral density in Figure 1 we got the two graphs in Figures 2–3 in agreement with theory. In the spectral density in Figure 4, however, the results look different, as can be seen in Figure 5. Here the efficiency of m_{LS} decreases and quite fast when n ranges from 10 to 50. The behavior of the C_ν as given in Figures 6–8.

These are only a few of the Calcomp graphs studied, but they may be sufficient to illustrate why we were led to the hypothesis to be described below.

It should be mentioned first that the computing was done in FORTRAN using the Calcomp plotter offline. It took a few weeks to complete. This was five years ago. If we had met the same problem tody we would have used APL and done the plotting online; this can be done in a few days, so that the time scale would be drastically changed.

The conditions for asymptotic efficiency of m_{LS} are of two types. One of them makes assumptions about the regularity properties of the spectral density. This is introduced to simplify the analytic treatment and can be weakened. Actually, already the first proof was quite general in this regard but it was also a

FIGURE 1

FIGURE 2

FIGURE 3

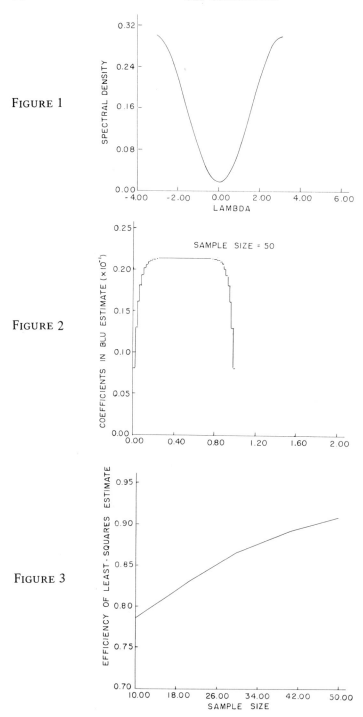

FIGURE 4

FIGURE 5

FIGURE 6

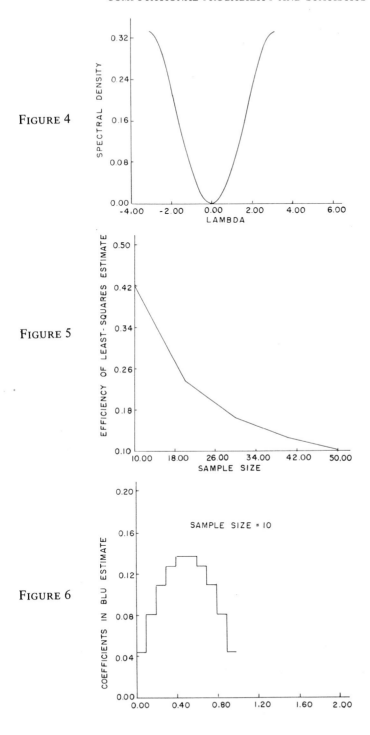

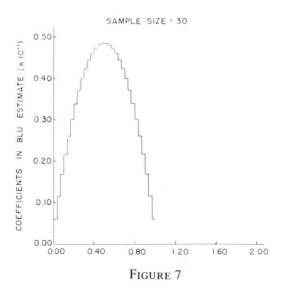

FIGURE 7

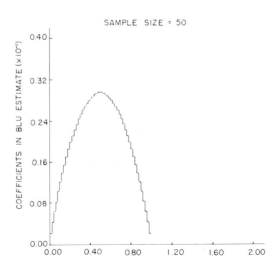

FIGURE 8

good deal more complicated than proofs that were suggested later on. The second type of condition is of more peculiar type: It assumes that the spectral density is bounded away from zero. It is embarrassing to have to make this assumption and some attempts have been made to remove it. A closer scrutiny of the problem shows that it is enough to assume that f is bounded away from zero in a neighborhood of the frequency zero, but it had not been possible to push the result further than this.

It should be pointed out that our lack of understanding of the role played by this condition could have substantial practical consequences. This would be so if the asymptotic efficiency of m_{LS} were small as opposed to what it would be if the theorem were applicable. In a real-life situation we seldom have precise information about f but we may know something about its qualitative behavior. This knowledge should then be used to get a better estimate that does not waste as much of the data. This problem does not seem to have been discussed in the literature.

The spectral density in Figure 4 has indeed a zero at $\lambda = 0$ which is, of course, why we chose it. Now looking at the graphs of C_ν one notices almost immediately that the shape appears to be parabolic. *This was actually observed in several other cases and we were led to study the hypothesis that an asymptotically efficient estimate could be obtained by choosing a parabolic form*

$$(6) \qquad C_\nu = A_n + B_n \nu + C_n \nu^2, \qquad \nu = 1, 2, \cdots, n.$$

Once the hypothesis had been formulated we could attempt to prove it by analytic means. We shall not give the proof here, and give only one of the results.

THEOREM. *Let f be continuous and positive except at $\lambda = 0$, where*

$$(7) \qquad \lim_{\lambda \to 0} f(\lambda)/\lambda^2 > 0.$$

Then the estimate m with the coefficients

$$(8) \qquad C_\nu = \frac{6}{n + 2} \frac{\nu}{n} \left(1 - \frac{\nu}{n + 1}\right)$$

is asymptotically efficient.

Using the technique of the proof we could also get the asymptotic form of $\mathrm{Var}(\hat{m})$ in closed form.

If the parabolic estimate m is used when $f(0) > 0$ it turns out that we do not lose very much. Indeed, its asymptotic efficiency is then equal to $5/6$.

Another strange feature that was observed in the numerical experiment was the behavior of the C_ν of the BLU estimate graphed in Figure 9. This was computed with the density f in Figure 10 and there is certainly nothing apparently pathological about it. The behavior of C_ν would be expected to be that of an almost constant sequence, not the wildly fluctuating one in Figure 9. A moment's reflection will suffice to realize that the asymptotic efficiency associated with $C_\nu' = 1/n$ does not imply that the sequence C_ν of the BLU estimate

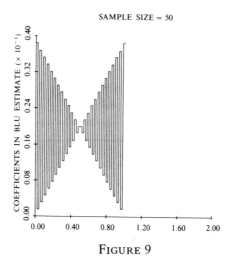

SAMPLE SIZE = 50

FIGURE 9

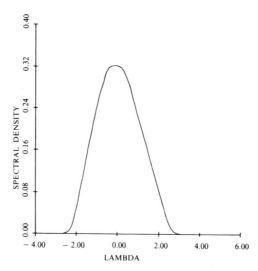

FIGURE 10

is asymptotically equal to that of the C'_ν: The theorem states only that the *corresponding variances* are asymptotically equal. It would be interesting, however, to pursue this further analytically and get a better insight into this lack of stability; this can no doubt be done and seems well motivated by the numerical experiment.

As another example, let G be a topological group and P a probability measure defined on the Borel sets of G. Take a sample of independent observations of size n from G according to the probability law P and denote the resulting group elements $g_1, g_2, g_3, \cdots, g_n$. Form the product

$$(9) \qquad \gamma_n = g_1 g_2 g_3 \cdots g_n.$$

A problem that has received a good deal of attention in recent years is the following one. The stochastic group element γ_n has some probability distribution P_n that can be viewed as the nth convolution power of P:

$$(10) \qquad P_n = P^{n*}.$$

How does P_n behave asymptotically for large n? Are there analogues to the probabilistic limit theorems on the real line or in other Euclidean vector spaces?

The state of our current knowledge depends very much on the structure of the group G. If G is compact the problem has been well penetrated and a number of informative theorems have been proved. For commutative, locally compact G we also know a good deal. When we proceed to groups which are locally compact but not commutative, our knowledge is rather scanty. *The basic normalization problem has not been settled*: How should we normalize γ_n in (9) in order to arrive at nontrivial probabilistic limit theorems? How should it be transformed by a one-to-one mapping into some other space? In a few special cases we know how to do it, but not in general.

In view of this it may appear as too early to turn to groups which are not locally compact. There are two reasons for doing this, however.

First, it may be that, again, we can deal with some special case that will tell something about what possibilities exist and what probabilistic behavior we should be prepared to meet in the absence of local compactness.

Second, a practical motivation arose from some work in pattern analysis. A space X that we shall refer to as the *background space* is mapped onto itself by a random function ϕ called the deformation:

$$(11) \qquad \phi: X \to X.$$

Very often ϕ is one-to-one and has some continuity property in probability. We are interested in what happens when ϕ is iterated:

$$(12) \qquad \begin{aligned} \psi_1(x) &= \phi_1(x), \\ \psi_{i+1}(x) &= \phi_{i+1}[\psi_i(x)], \qquad i = 1, 2, 3, \cdots, \end{aligned}$$

where $\{\phi_i\}$ is a sequence of random functions, independent of each other and with the same probability measure P on some function space Φ. This leads obviously to the problem described above.

In most practical cases X is a Euclidean vector space or a subset of one. We shall look at the case $X = [0, 1]$ and

(13)
$$\begin{cases} \phi \text{ is nondecreasing,} \\ 0 \leqslant \phi(x) \leqslant 1, \\ \phi(0) = 0, \quad \phi(1) = 1, \end{cases}$$

as minimum requirements. We shall make it a bit stronger by assuming ϕ to be continuous in probability. Sometimes we have also assumed that ϕ is strictly increasing so that we get a group, not just a semigroup.

We shall use two functions to give a crude characterization of the deformation: the mean value function $m(x)$ and the variance function $V(x)$:

(14)
$$m(x) = E\,[\phi(x)], \qquad V(x) = \text{Var}\,[\phi(x)].$$

If the deformation has a tendency to the left in the sense that $m(x) < x, 0 < x < 1$, we call the deformation *left-systematic*, and similarly right-systematic. If $m(x) \equiv x$, which is a practically important case, the deformation is called *fair*.

This is not the right place to describe the study of random deformations of a background space in any depth. We shall just indicate how we used the computer to help us look at convolution powers on this group and how this aided us in suggesting some theorems. At this point we should stress the psychological desirability of not just computing the convolution process but of *making the computational results intuitively available* by graphic representation or by other technical means. In this study this was done by crude APL plots and also by using Calcomp to get better graphic resolution.

A first question is: When do we reach equilibrium distributions and what form do they have? It is not difficult to prove that if the deformation is left- (right-) systematic, an equilibrium distribution must be concentrated on the two-point set $\{0, 1\}$. To shed some light on the fair deformations we generated a series of Calcomp plots, some of which are given in Figures 11–15 where ψ_n is plotted against x for $n = 10, 20, 30, 40, 50$. Other graphs had the same appearance: ψ_n tends to a step function with a single step and the location of the step varies from sequence to sequence but not within a sequence. This led us to the following conjecture.

THEOREM. *If ϕ is a fair deformation with a continuous $V(x)$ vanishing only at $x = 0, 1$, there exists a stochastic variable ξ such that*

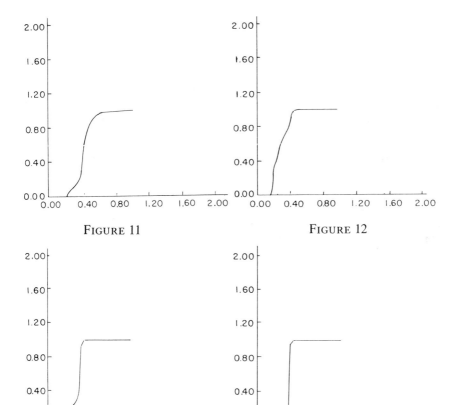

FIGURE 11   FIGURE 12

FIGURE 13   FIGURE 14

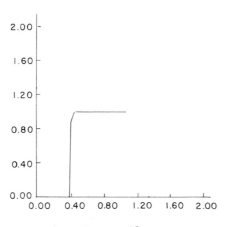

FIGURE 15

$$(15) \qquad \lim_{n \to \infty} \psi_n(x) = \begin{cases} 0 & if \ x < \xi, \\ 1 & if \ x > \xi, \end{cases}$$

and where ξ has the rectangular distribution $R(0, 1)$.

Once formulated, the conjecture was not difficult to prove; we just had not asked the right question until the computer experiment showed us the way. Indeed, at first it was something of a shock to see the way the graphs looked with their counterintuitive shapes.

If a (univariate) equilibrium distribution exists for $\psi_n(x)$ for any (fixed) x, the question arises: What can we say about bivariate (and higher-dimensional) equilibrium distributions? Some APL plots were helpful: We show two in Figures 16 and 17. It is apparent that the points tend to cluster on the $45°$ diagonal. It suggested the following theorem.

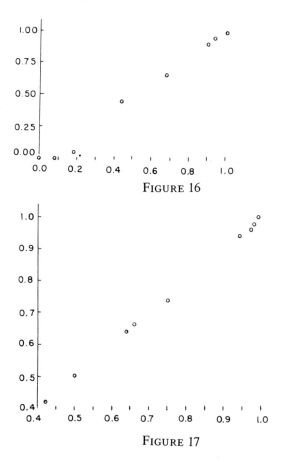

FIGURE 16

FIGURE 17

THEOREM. *If the univariate limit distribution is unique and does not depend on* x, *then any bivariate limit distribution is singular and has all its mass on the diagonal of the unit square.*

The proof is quite simple. Several other computer experiments were carried out but will not be discussed here.

We have only let the computer help us in making guesses here. These guesses have then been verified (or discarded, occasionally) by using deductive reasoning and the standard analytical tools.

Some caution is needed considering the interpretation of topological concepts, such as those occurring above, on the computer. It could be argued that since the machine is a finite device, it is meaningless to implement concepts like continuity and compactness on it. This is true only on a superficial level. The argument misses the point that qualitative properties—without complete correspondence to the computer—can be expressed quantitatively in discrete terms resulting in a meaningful correspondence. One could go still further and say that a property, topological or not, that appears in a theorem is practically irrelevant if it cannot be formulated into a meaningful property expressed in discrete terms. This translation is not a trivial thing to do, however; it takes experience and intuitive insight.

A teaching and research activity of this type does not operate in a technological vacuum. Attention must be paid to the technical aspects; otherwise, the important part of it will be overshadowed by software and hardware problems, drawing the attention away from the mathematics.

Without access to reliable interactive computing equipment it would be difficult to carry out the program. Some sort of graphic output device is also needed, but it need not be sophisticated and powerful; a simple plotter is enough, and even typewritten plots may be adequate.

As far as programming languages are concerned, there was never any doubt in our minds that APL is by far the superior one for this purpose. High-dimensional arrays are easy to handle, the primitive operators are well chosen, the whole language is close to mathematical language, and the debugging facilities in the 360-implementation are helpful. It is true that the absence of a compiler makes certain computations expensive and bulk I/O was not available until recently. On the whole, however, it is an excellent language.

We set out to see if, and in what way, advances in computer technology have influenced our attitude toward the study and teaching of mathematical statistics. It would be premature to project any drastic change in the way the community of mathematical statisticians approach these problems, but we have explored a number of cases where the impact of the modern computer, and especially time-sharing capacity, could and should be felt.

Access to effective time-sharing facilities is not as widespread as one might wish, but the situation is improving. This will happen gradually and will be accompanied by a more general appreciation of its consequences for probability and statistics. Adding these powerful resources to the analytic methods that we have been using traditionally will amplify the effect of the latter, increasing the scope of mathematical statistics.

BROWN UNIVERSITY

Proceedings of Symposia in Applied Mathematics
Volume 20
1974

AN INTRODUCTION TO SOME CURRENT RESEARCH
IN NUMERICAL COMPUTATIONAL COMPLEXITY [1]

BY

J. F. TRAUB

This is a Conference on the Influence of Computing on Mathematical Research and Education, I want to talk about a particular area of numerical mathematics, *numerical computational complexity*, which has a very large intersection with mathematics. On the one hand, it requires the development of new mathematical techniques while, on the other hand, certain of the questions it tries to answer are primarily mathematical.

In this paper I will focus on research rather than educational issues. However, I will mention that some nontrivial results and questions can be formulated so as to be accessible to both undergraduates and graduate students. This is at least partially the case because the field is so new.

I will not attempt a survey of current work in computational complexity; rather, I will restrict myself to problems and algorithms from numerical mathematics. Within numerical computational complexity, I will try to give you the *flavor* of some of the recent work and state a few open problems. I will draw on work done at Carnegie-Mellon University for many of my examples.

A number of the papers I will cite were presented at a Symposium on Complexity of Sequential and Parallel Numerical Algorithms at Carnegie-Mellon University in May, 1973. Incidentally, there have been at least four symposia in the United States from April to July, 1973, devoted entirely or in major part to computational complexity.

I will not give a formal definition of complexity here. Instead I will begin

---

AMS (MOS) subject classifications (1970). Primary 68A20, 68A10, 68–02, 65H05.

[1]This research was supported in part by the National Science Foundation under grant GJ-32111 and the office of Naval Research under contract N0014-67-A-0314-0010, NR 044-422.

with a particular problem, that of matrix multiplication, to illustrate some basic ideas.

Matrix multiplication. Let A, B be (n, n) matrices and let $C = AB$. The definition of matrix product gives us the following algorithm for forming the n^2 elements of C—take the scalar product of rows of A with columns of B. This algorithm takes $O(n^3)$ multiplications and $O(n^3)$ additions. Until about five years ago no one asked if there might be better algorithms. Then Winograd [27] showed that matrix multiplication could be done in $\frac{1}{2}n^3 + O(n^2)$ multiplications and Strassen [22] gave an algorithm which required fewer than $O(n^3)$ arithmetic operations. Consider, in particular, the problem of multiplying two $(2, 2)$ matrices. Classically, this requires eight scalar multiplications. Strassen showed how to multiply the matrices in seven scalar multiplications. Furthermore, his algorithm does not depend on commutativity of multiplication of the elements and can therefore be applied to partitioned matrices. He showed that two (n, n) matrices could be multiplied in $O(n^{\log_2 7})$ *arithmetic operations.* Since $\log_2 7 \sim 2.81$, the multiplication of two matrices can be done in $O(n^{2.81})$ arithmetic operations.

Now for very large matrices the fact that matrix multiplication can be done in $O(n^{2.81})$ arithmetic operations might be of interest. It is, however, the research engendered by this result and its theoretical consequence which is of greater interest than the practical applications.

Can we do better than $O(n^{2.81})$? Not by using 2 by 2 matrices since it has been established that seven multiplications are optimal. What about using 3 by 3 matrices? The minimum number of multiplications for multiplying two 3 by 3 matrices is open.

Can we say anything in general about the minimum number of arithmetic operations needed to multiply any n by n matrices? Since the problem has $2n^2$ inputs and n^2 outputs, $O(n^2)$ must be a lower bound on the number of arithmetics. Now $O(n^2)$ is linear in the number of inputs and outputs. Is there a nonlinear lower bound? We do not know. Our present state of knowledge is

$$O(n^2) \leqslant \text{number of arithmetic operations} \leqslant O(n^{\log_2 7}).$$

A survey of what is known is given by Borodin [3].

I will use the matrix multiplication problem to illustrate some general features of *algebraic complexity.* There is a parameter which is a measure of the problem size. In the matrix problem this is the order of the matrix. There are one or more fundamental operations. In the matrix multiplication case this is the number of scalar multiplications or the number of scalar arithmetic operations. There is an algorithm for solving the problem in a finite number of fundamental operations. This algorithm gives us an upper bound on the difficulty of the problem. We also want lower bounds and preferably tight lower bounds. For the matrix problem we saw $O(n^2)$ was a lower bound linear in the number of inputs/outputs. We regard such a lower bound as trivial; we want nonlinear lower bounds. Few such lower bounds are known today.

We call the difficulty of a problem its computational complexity. We also refer to the complexity of a particular algorithm. How the difficulty is measured depends on what we designate as the fundamental operation or operations. For example, for a parallel computer it is usually the time rather than the number of operations that we try to optimize. I will return to this later.

The area I am discussing is often called concrete or specific complexity to distinguish it from the abstract complexity theory of Hartmanis, Blum and others. John Hopcroft wishes to reserve the name computational complexity to lower bound results. He feels that upper bounds represent computational *simplicity*. From a purely theoretical point of view this may be true. From a pragmatic point of view, good upper bounds provided by algorithms are very useful.

I find it convenient to divide computational complexity into two parts: *algebraic computational complexity* and *analytic computational complexity*.

Algebraic computational complexity deals with problems for which there are finite algorithms. Analytic computational complexity deals with problems for which there are no finite algorithms. I will discuss analytic complexity later. Here I want to add several more examples of algebraic complexity to the matrix problem discussed above.

The following problems deal with nth degree polynomials and they all take $O(n^2)$ operations if classical algorithms are used:

Multiplication of two polynomials.

Division by a polynomial of degree $n/2$.

Evaluation of a polynomial at n points.

Interpolation at $n + 1$ points.

Evaluation of a polynomial and all its derivatives at one point.

Recently, algorithms have been developed for doing the first two problems in $O(n \log n)$ arithmetic operations and the last three in $O(n \log^2 n)$ arithmetic operations. Results here are due to Borodin, Horowitz, Kung, Strassen, and others. (See Borodin [3].) Using arguments from algebraic geometry, Strassen [23] has shown that the interpolation problem requires $O(n \log n)$ multiplications. This is a nonlinear lower bound. Thus for the interpolation problem we have tight bounds from above and below:

$$O(n \log n) \leqslant \text{ number of arithmetic operations } \leqslant O(n \log^2 n).$$

The results above are asymptotic. But most problems arising in practice are for small n. For one of these problems, the evaluation of a polynomial and its derivatives at a point, we do have a result which is better than the classical algorithm for all n. The classical algorithm is the iterated Horner method which uses $\frac{1}{2}n(n + 1)$ multiplications and $\frac{1}{2}n(n + 1)$ additions. Shaw and Traub [19] have introduced an algorithm which requires $3n - 2$ multiplications and divisions and $\frac{1}{2}n(n + 1)$ additions. Thus the number of multiplications and divisions is linear rather than quadratic. There are a number of open questions including:

(1) Is there an algorithm using only a linear number of additions?

(2) It is easy to show at least $n + 1$ multiplications are required. An upper bound is $3n - 2$ multiplications and divisions. Can these bounds be tightened?

(3) How many multiplications are required if divisions are not permitted?

An area which has seen a great deal of recent activity is the fast solution of linear systems, particularly of sparse systems arising, for example, from the discretization of partial differential equations. Recent surveys are given by Bunch [7] and Birkhoff and George [2].

One particular problem is the solution of a linear system whose matrix M is n^2 by n^2 and in block tridiagonal form, $M = [-I, T, -I]$, where T is an n by n tridiagonal matrix and I is the n by n identity. There are therefore $O(n^2)$ nonzero elements. Bank, Birkhoff, and Rose [1] give an $O(n^2)$ algorithm. This gives an upper bound linear in the number of nonzero elements and we cannot hope for better than that. However, stability is crucial in this area and this algorithm is not stable. P. Swartztrauber (private communication) has developed a stable $O(n^2 \log \log n)$ algorithm.

So far, I have confined myself to problems for which there are finite algorithms. I used the adjective algebraic to characterize the complexity of problems in this area. I will now turn to an area where there are no finite algorithms. I will refer to this area as analytic computational complexity.

Recent work in this area which I will not describe further in this paper includes Eisenstat and Schultz [8] on the complexity of partial differential equations, Rice [18] on approximation, Brent [5] on systems of nonlinear equations, and Brent, Winograd and Wolfe [6] on optimal iteration.

As a simple first illustration of a problem in analytic complexity, let us consider the calculation of α, $\alpha = A^{\frac{1}{2}}$. A well-known method for calculating α is as the limit of the sequence $\{x_i\}$ defined by

$$(1) \hspace{4cm} x_{i+1} = \frac{1}{2}(x_i + A/x_i).$$

This is Newton iteration applied to $f = x^2 - A$. We can ask about the complexity of $A^{\frac{1}{2}}$. An upper bound is obtained from the complexity of the particular algorithm specified by (1). We will return to this problem after we have developed some tools.

Let x_0 be given and let $x_{i+1} = \varphi(x_i)$ be a scalar iteration with $x_i \to \alpha$. Thus φ, x_0 defines an algorithm for computing α. What is the "efficiency" of this algorithm? I first have to introduce the basic concept of order. Let $p = p(\varphi)$ denote the order of φ. Roughly speaking

$$x_{i+1} - \alpha = O[(x_i - \alpha)^p].$$

Let $C(\varphi)$ denote the "cost" of calculating x_{i+1} from x_i. We will give examples of various costs later. They have the critical property that $C(\varphi \circ \varphi) = 2C(\varphi)$. Define the efficiency of φ by

$$e(\varphi) = (\log p(\varphi))/C(\varphi).$$

All logarithms are to base two.

This efficiency has the following properties:

(1) It is inversely proportional to the total cost of estimating α to within ϵ, for a small $\epsilon > 0$.

(2) It is invariant under self-composition. That is, $e(\varphi \circ \varphi) = e(\varphi)$.

There are various quantities that can be used for C. For example, we can take the following:

(1) $C = M =$ number of multiplications or divisions to compute x_{i+1} from x_i.

(2) $C = \overline{M}$. This is the same as M except that multiplication by constants is not counted.

Let

$$e = (\log p)/M, \quad \overline{e} = (\log p)/\overline{M}.$$

Clearly $e \leqslant \overline{e}$. Kung [10] proved that if φ is *any rational function*, then $\overline{e} \leqslant 1$ and this bound is sharp. Therefore $e \leqslant 1$. Since a computer performs rational operations, this is not restrictive. Next, Kung [11] asked for what iterations is $e = 1$ and for what iterations is $\overline{e} = 1$? He completely settled this question as follows:

If $e = 1$, then α is rational.

If $\overline{e} = 1$, then α is rational or quadratic irrational.

Thus only easy problems have optimal efficiency.

The above result concerns optimal φ over all problems α. Now let us consider a particular α. To fix ideas, let $\alpha = A^{\frac{1}{2}}$. What's the best iteration for calculating $A^{\frac{1}{2}}$? The most widely known algorithm is

$$x_{i+1} = \tfrac{1}{2}(x_i + A/x_i).$$

Then $p = 2, \overline{M} = 1, M = 2$, and therefore

$$\overline{e} = (\log 2)/M = 1, \quad e = (\log 2)/M = \tfrac{1}{2}.$$

Thus, in the $\overline{e}$ measure, Newton-Raphson iteration is optimal. This was first pointed out by Paterson [17]. However, in the e measure, the question is open. What is the optimal iteration for computing a quadratic irrational in the multiplicative efficiency measure e? One can ask similar questions for classes of mathematical problems.

Observe that the cost in the above development is limited to multiplications. The reason for this is technical. Multiplications can effect the degree growth of a polynomial. Additions do not cause polynomial degree to increase and therefore analysis based on degree growth cannot be used for additions. Morgenstern [15] has made recent progress on using other growth arguments for addition.

I am going to change gears here and discuss approximating a zero α of a scalar function f by a sequence of iterations x_i. This problem is a prototype of many problems of applied mathematics where we seek a zero of an operator equation (Traub [24]). Examples are the numerical solution of integral equations and partial differential equations. With the exception of recent work by Brent [5] and Wozniakowski [28] on finding a zero of a vector-valued function, that is solving a system of nonlinear equations, almost all work has dealt with zeros of scalar functions.

Kung and Traub [13] introduced a new measure for the efficiency of an algorithm φ with respect to a problem f (see also Traub [26]). I do not want to get into technical details here other than to mention that they include the *combinatorial* cost of φ, which is the number of arithmetic operations used by an iteration even if the evaluation of f and its derivatives were free. It turns out to be crucial to include the combinatorial cost.

There is a major open conjecture in the area of iteration algorithms for calculating zeros of a function f. I remind you that, roughly speaking, the order of an iteration φ is a number p such that

$$x_{i+1} - \alpha = O((x_i - \alpha)^p).$$

Let the total number of evaluations of f or its derivatives that φ uses in going from x_i to x_{i+1} be n.

Kung and Traub [12] have constructed an algorithm using n evaluations which is of order 2^{n-1} and they *conjecture* this to be optimal. That is, the order of any iteration without memory (iteration is precisely defined by Kung and Traub [12]) based on n evaluations is of order at most 2^{n-1} for every positive integer n. The conjecture has been proven for $n = 1, 2$ (Kung and Traub [14]).

Up to now we have implicitly assumed we were dealing with a *sequential computer*. Thus the cost associated with an algorithm has been in terms of fundamental operations, such as the number of arithmetics. On a *parallel machine*, the cost associated with an algorithm is *time* and this is the quantity we optimize.

To a first approximation you can think of a parallel computer as consisting of a set of arithmetic units which you can use simultaneously. I emphasize this is a first approximation only. There are major differences in the parallel and vector computers now becoming available. Furthermore, when preparing programs for these machines, matters of data handling and data structure become paramount.

I think algorithms for parallel computers are one of the most interesting new research areas in numerical mathematics for a number of reasons:

(1) From a practical point of view, because the parallel machines are coming. Machines which should be available within a year or so including Carnegie-Mellon University's Multi-Mini Computer, CDC STAR, Burrough's ILLIAC 4, Texas Instruments' ASC. We need algorithms to put on them. In general, the best classical algorithms will not be very good on parallel machines.

(2) From a theoretical point of view, because constructive mathematics has been implicitly sequential. The problems raised by parallelism raise a whole new world of interesting mathematical issues.

On a parallel machine we will be concerned with the *time* taken by an algorithm. One basic quantity to be analyzed is the parallel speed-up ratio defined as follows. For a problem of size m, the speed-up is defined as

$$S(m, k) = \frac{\text{optimal computation time on a one-processor computer}}{\text{optimal computation time on a } k\text{-processor computer}}.$$

In general we do not know the optimal times and therefore can only get estimates and bounds on $S(m, k)$.

Stone [20] considers the maximal speed-up which can be achieved for various problems. Trivially, $S(m, m) \leqslant m$. A simple example of a problem with linear speed-up is the addition of two matrices of order m. On a sequential machine this takes m^2 addition time. On a parallel machine with m^2 processors this can be done in one addition time. Hence the speed-up is linear in the number of processors. Problems such as linear recurrence investigated by Stone [21] and Heller [9], poly-

nomial evaluation (Munro and Paterson [16]), rational evaluation (Brent [4])
have sped-up at least $m/(\log m)$ which is close to linear. The numerical solution
of partial differential equations also seems amenable to parallel computation. The
solution of tridiagonal systems, which arise in the discretization of partial differential equations, is analyzed by Traub [25]. The study of parallel algorithms is in an
embryonic state. I believe it will be one of the major research areas in numerical
mathematics in the next decade.

I have tried to give you a taste of some of the current research in numerical
computational complexity. I would like in conclusion to summarize some of my
reasons for studying complexity:

(1) To construct good new algorithms. There are, however, many components besides complexity to be considered in constructing a good algorithm.

(2) To filter out bad algorithms.

(3) To create a theory of algorithms which will establish theoretical limits on
computation.

(4) To investigate the intrinsic difficulty of mathematical problems.

Acknowledgment. I would like to thank H. T. Kung for his comments on
this paper.

BIBLIOGRAPHY

1. R. Bank, G. Birkhoff and D. J. Rose, *An $O(N^2)$ method for solving constant coefficient boundary value problems in two dimensions*, Report, Harvard University, Cambridge,
Mass., 1973.

2. G. Birkhoff and A. George, *Elimination by nested dissection*, Complexity of Sequential and Parallel Numerical Algorithms, Academic Press, New York, 1973, pp. 221–270.

3. A. Borodin, *On the number of arithmetics required to compute certain functions—
Circa May* 1973, Complexity of Sequential and Parallel Numerical Algorithms, Academic Press,
New York, 1973, pp. 149–180.

4. R. Brent, *The parallel evaluation of arithmetic expressions in logarithmic time*, Complexity of Sequential and Parallel Numerical Algorithms, Academic Press, New York, 1973,
pp. 83–102.

5. ———, *Some efficient algorithms for solving systems of nonlinear equations*, SIAM
J. Numer. Anal. 10 (1973), 327–344.

6. R. Brent, S. Winograd and P. Wolfe, *Optimal iterative processes for root-finding*,
Numer. Math. 20 (1973), 327–341.

7. J. R. Bunch, *Complexity of sparse elimination*, Complexity of Sequential and Parallel
Numerical Algorithms, Academic Press, New York, 1973, pp. 197–220.

8. S. C. Eisenstat and M. H. Schultz, *The complexity of partial differential equations*,
Complexity of Sequential and Parallel Numerical Algorithms, Academic Press, New York, 1973,
pp. 271–282.

9. D. Heller, *A determinant theorem with applications to parallel algorithms*, SIAM J.
Numer. Anal. (to appear). (Also available as a CMU Computer Science Department report.)

10. H. T. Kung, *A bound on the multiplication efficiency of iteration*, JCSS **7** (1973), 334–342. (Also available as a CMU Computer Science Department report.)

11. ――――, *The computational complexity of algebraic numbers*, SIAM J. Numer. Anal. (to appear). (Also available as a CMU Computer Science Department report.)

12. H. T. Kung and J. F. Traub, *Optimal order of one-point and multipoint iteration*, J. ACM (to appear). (Also available as a CMU Computer Science Department report.)

13. ――――, *Computational complexity of one-point and multipoint iteration*, SIAM-AMS Proc., vol. 7, Amer. Math. Soc., Providence, R. I., 1974 (to appear). (Also available as a CMU Computer Science Department report.)

14. ――――, *Optimal efficiency and order for iterations using two evaluations*, Report, Department of Computer Science, Carnegie-Mellon University, 1973.

15. J. Morgenstern, *Note on a lower bound of the linear complexity of the fast Fourier transformation*, J. ACM, April 1973.

16. L. Munro and M. Paterson, *Optimal algorithms for parallel polynomial evaluation*, J. Comput. System Sci. **7** (1973), 189–198.

17. M. S. Paterson, *Efficient iterations for algebraic numbers*, Complexity of Computer Computations (edited by R. E. Miller and J. W. Thatcher), Plenum Press, New York, 1972, pp. 41–52.

18. J. R. Rice, *On the computational complexity of approximation operators*, Report Purdue University, Fort Wayne, Ind., 1973.

19. M. Shaw and J. F. Traub, *On the number of multiplications for the evaluation of a polynomial and some of its derivatives*, J. ACM (to appear). (Also available as a CMU Computer Science Department report.)

20. H. S. Stone, *Problems of parallel computation*, Complexity of Sequential and Parallel Numerical Algorithms, Academic Press, New York, 1973, pp. 1–16.

21. ――――, *An efficient parallel algorithm for the solution of a tridiagonal system of equations*, J. ACM **20** (1973), 27–38.

22. V. Strassen, *Gaussian elimination is not optimal*, Numer. Math. **13** (1969), 354–356. MR 40 #2223.

23. ――――, *Die Berechnungskomplexitat von elementarsymmetrischen Funktionen und von Interpolationskoeffizienten*, Numer. Math. **20** (1973), 238–251.

24. J. F. Traub, *Computational complexity of iterative processes*, SIAM J. Comput. **1** (1972), 167–179. (Also available as a CMU Computer Science report.)

25. ――――, *Iterative solution of tridiagonal systems on parallel or vector computers*, Complexity of Sequential and Parallel Numerical Algorithms, Academic Press, New York, 1973, pp. 49–82. (Also available as a CMU Computer Science report.)

26. ――――, *Theory of optimal algorithms*, Proc. Conference on Software for Numerical Mathematics, Loughborough, England, 1973 (to appear). (Also available as a CMU Computer Science report.)

27. S. Winograd, *A new algorithm for inner-product*, IEEE Trans. C-17 (1968), 693–694.

28. H. Wozniakowski, *Maximal stationary iterative methods for the solution of operator equations*, SIAM J. Numer. Anal. (to appear).

CARNEGIE-MELLON UNIVERSITY

Proceedings of Symposia in Applied Mathematics
Volume 20
1974

APPLIED MATHEMATICS AND COMPUTING [1]

BY

PETER D. LAX

1. **Introduction.** It is a truism that the advent of powerful computing machines has revolutionized applied mathematics, its classical branches as well as the modern ones. In this paper I will describe some new developments in one corner of mathematical physics, the approximate solution of the initial value problem for the partial differential equations of mathematical physics. Characteristic of these new developments is the employment of ingenious algorithms for the evaluation of the operators in the approximation scheme. The feeling is unavoidable that there are tricks not yet discovered which could further reduce the arithmetic labor needed to find approximations within given error bounds; if so, we are indeed very far from an adequate theory of computational complexity for problems of this kind.

The initial value problems we consider here are of the form

$$(1.1) \qquad u_t = Cu, \quad u(0) \text{ given.}$$

We shall take C to be a linear partial differential operator independent of t; in this case the solution to (1.1) can be written symbolically as

$$(1.2) \qquad u(t) = e^{Ct}u(0).$$

The approximation schemes we shall consider will be, with the exception of the last one, of the following form:

Given some small quantity Δt we construct an operator S such that $Su(0)$ approximates $u(\Delta t)$. Then we choose

$$(1.3) \qquad S^n u(0), \quad n\Delta t = t,$$

AMS (MOS) subject classifications (1970). Primary 65M10, 65M99.

[1] This research was supported by the U. S. Atomic Energy Commission, contract AT(11-1)−3077.

as an approximation to $u(t)$. To decide how good a particular proposed scheme is
we need realistic answers to the following questions:

(i) How close is $S^n u(0)$ to $u(t)$?

(ii) What is the computational cost of evaluating $S^n u(0)$?

In §2 we describe one aspect of question (i), and in §3 some answers to
(ii) are discussed.

Some of the schemes and algorithmic devices described in §3 are applicable
to nonlinear equations as well. But, generally nonlinear problems have peculiar
features which have no counterpart in the linear case, and these present further
difficulties for the construction of efficient approximation schemes. Much recent
work in computational fluid dynamics has been aimed at overcoming these difficulties; in some cases the nonlinear character of the problem has been used to
advantage. I hope to survey these developments in another paper.

I recommend the field of numerical solutions of partial differential equations to young mathematicians looking for a research area. The problems are of
great practical importance, and a clever idea can pay off handsomely.

2. **Stability of numerical schemes.** The accuracy of a numerical scheme
depends not only on how good an approximation $Su(0)$ is to $u(\Delta t)$, but also on how
much an initial discrepancy is magnified during the course of the calculation. If
it is not magnified at all, or only moderately, the scheme is called *stable*; in this
section we discuss some mathematical problems concerning stability.

An initial discrepancy d contributes $S^n d$ to the approximate solution
at time $t = n\Delta t$. Stability means that the operator S^n does not magnify d too
much, i.e., that the norms of these operators are uniformly bounded:

(2.1) $|S^n| \leqslant K$ for all n,

K independent of n and Δt. The question we shall discuss here is how to
decide that (2.1) holds for some given operator S proposed in some approximation scheme.

An obvious sufficient condition is that

(2.2) $|S| \leqslant 1$;

clearly, (2.2) implies (2.1), with $K = 1$. When (2.2) is satisfied, the scheme is
called *strongly stable*. Such schemes are often associated with differential equations where energy is conserved, or decreases with time. But in many cases of
practical interest condition (2.2) does not hold, yet the method is stable. We describe now some such sophisticated conditions for stability.

Suppose C is a linear partial differential operator with constant coefficients which are square matrices, and the operator S is a difference operator with constant matrix coefficients. In this case Fourier transformation changes the action of S to multiplication by a matrix valued function M; to decide whether the powers of S are bounded we have to know whether the powers of M are bounded.

An obvious necessary condition for the uniform boundedness of M^n for all n is that none of the eigenvalues of M exceed 1. There are a number of sufficient conditions; the most powerful and most useful for stability theory is the following criterion due to H. Kreiss:

$|M^n| \leq K$ *for all* n *iff the resolvent of* M *satisfies*

$$|(I - zM)^{-1}| \leq k/(1 - |z|)$$

for all z *with* $|z| < 1$. *The constant* K *depends on* k *and on the order of* M.

For a proof and further discussion see Richtmyer and Morton's book [5] on the initial value problem.

Another useful sufficient condition for the boundedness of M^n, due to the author and B. Wendroff (see [3]), relates power boundedness to the *numerical range*:

If $|(w, Mw)| \leq (w, w)$ *for all complex vectors* w, *then, for all* n, $|M^n| \leq K$ *where* K *depends only on the order of* M.

It was conjectured by Halmos and proved by C. Berger, and subsequently by C. Pearcy, that the constant K above can be taken as 2. For details, see e.g. [2].

3. Fast algorithms. In this section we present a number of numerical schemes each of which is distinguished by employing an algorithm which makes the evaluation of the operations less costly than it seems at first.

The computational cost of evaluating S^n is (at most) n times the cost of a single application of S. As we shall see, S requires a number of multiplications and divisions by matrices of fairly high order. Savings can come from employing only such matrices for which multiplication and inversion can be performed efficiently. A survey of such methods is presented in the volume [6].

Our first example is an old one: the solution of the initial value problem for parabolic equations in one space variable of which the heat equation is typical:

(3.1) $u_t = u_{xx}.$

In the most direct, so-called explicit method, one replaces the t derivative by a time difference centered at time $t + \Delta t/2$ and the space derivative by a symmetric difference evaluated at the old time t. The resulting difference scheme has two well-nigh fatal flaws: It is not very accurate, and it is unstable unless Δt is taken to be less than $\frac{1}{2}(\Delta x)^2$, where Δx is the size of the spatial mesh. For realistic choices of Δx the resulting Δt is so small that the number n of time steps is unacceptably large. A way of overcoming both defects is to approximate u_{xx} by an average of symmetric differences evaluated at the old time and the new time. Since in the second scheme both time and space derivatives are approximated by a difference quotient centered at $x, t + \Delta t/2$, the second scheme is more accurate than the first. It is far less obvious but true that the second scheme is also more stable than the first; in fact the second scheme is unconditionally stable, no matter how large Δt is compared to Δx. On the other hand the second scheme is *implicit*, i.e., instead of giving explicitly the values of u at the new time it merely gives relations among the new values of u at various meshpoints; the extra labor of extracting the new values of u from these relations is the price we have to pay for the improvement in accuracy and stability.

How high is this price? The implicit relations are a system of linear equations of a particularly simple structure: Each relation links together values of $u(t + \Delta t)$ at 3 consecutive meshpoints. The matrix $\{a_{ij}\}$ describing such a system of equations is tridiagonal; i.e., the only nonzero entries a_{ij} are those with $|i - j| \leqslant 1$. It turns out that by using Gaussian elimination one can invert tridiagonal matrices cheaply; the number of arithmetic operations is proportional to the number of unknowns, which in this case is proportional to the number of meshpoints. See pp. 198–201 of [5] for details.

This is the place to point out that for computations done on a machine with parallel features—one with a pipeline-type arithmetic unit or one with many arithmetic units—the first, explicit method may be faster than the second, implicit method. For the explicit calculations can be carried out in parallel, whereas the operations for inverting a tridiagonal matrix are nested and therefore must be carried out serially one after another.

We turn now to the initial value problem for the heat equation in two space variables:

(3.2) $u_t = u_{xx} + u_{yy}.$

The explicit method suffers from the same defects as in the case of one space variables, so people have turned to implicit methods. The exact 2-dimensional analogue of the implicit method described above leads to a system of equations where each equation links together values u at 5 neighboring meshpoints; the matrix of such a system of equations cannot be inverted as economically as a tridiagonal matrix. Therefore the standard implicit method is far less practical for two space dimensions than it is in one dimension.

In the mid-fifties Peaceman and Rachford [4] invented an ingenious implicit scheme, called the *alternating direction method*, which is accurate, stable and involves the inversion only of tridiagonal matrices. The method employs alternatingly two different time steps; in the first phase the u_{xx} derivative is approximated by a difference quotient evaluated at the old time t, u_{yy} by a difference quotient evaluated at the new time $t + \Delta t$. In the second phase the roles of x and y are reversed. It is true, although not quite obvious, that this method is accurate and unconditionally stable. But quite clearly only tridiagonal matrices need be inverted since each of the two time steps treats only one of the variables implicitly.

The Russian school (see the book [8] by Yanenko) has championed an even more radical version of this idea, called the *method of fractional steps*. In this method the main time step is broken up into several phases. Here is how the method works on equation (3.2).

In the first phase ignore completely the term u_{xx} and replace u_{yy} by a difference quotient evaluated at time $t + \Delta t$; in the second phase the role of x and y is reversed. To analyze why and how well such a scheme works we put the kind of problem to which it can be applied into an abstract framework:

$$(3.3) \qquad u_t = (A + B)u,$$

A and B linear operators, independent of t. Then the exact solution can be put into the form

$$u(t) = e^{(A+B)t}u(0).$$

In particular, the operator which furnishes the exact solution at time Δt out of the initial data is

$$(3.4) \qquad e^{(A+B)\Delta t}.$$

Unless the operators A and B commute, the functional equation for the exponential function does not apply to operator arguments. That is, the product

(3.5) $e^{A\Delta t}e^{B\Delta t}$

is not equal to (3.4). However, the Taylor series expansion for the exponential
function shows that (3.5) differs from (3.4) by $\frac{1}{2}(\Delta t)^2[AB - BA] + O(\Delta t)^3$
and therefore (3.5) can be regarded as an approximation to (3.4). Finally we
approximate each of the factors $e^{A\Delta t}$ and $e^{B\Delta t}$ by operation S_A and S_B
and take

(3.6) $S = S_A S_B$

to be our approximation to (3.4). Formula (3.6) is the method of fractional
steps; clearly it is applicable to any number of fractions, not just 2. The great
advantage of the method of fractional steps is that the computational cost is
simply the sum of the computational costs of each fractional step. Another advantage is its stability; suppose that each fractional step is *strongly stable*, in the
sense that the norm of each operator S_A and S_B does not exceed 1. It then
follows that the norm of the product $S_A S_B$ is also $\leqslant 1$, which assures the
strong stability of the scheme based on (3.6).

        No matter how accurate approximations S_A and S_B are to $e^{A\Delta t}$ and
$e^{B\Delta t}$, the accuracy of the fractional step method is limited by the large discrepancy between (3.4) and (3.5). This can be remedied in several ways; interchange
the role of A and B, which turns (3.5) into

(3.5)' $e^{B\Delta t}e^{A\Delta t}$.

It is easy to show that although both (3.5) and (3.5)' differ from (3.4) by a term
of order $(\Delta t)^2$, their arithmetic mean approximates (3.4) with a discrepancy of
order $(\Delta t)^3$. So the modified fractional step method

(3.6)' $(S_A S_B + S_B S_A)/2$

is more accurate than (3.6).

        Formula (3.6)' takes twice as long to evaluate as (3.6). Strang has observed
that it is more advantageous to take the geometric rather than the arithmetic
mean of the fractional steps. It is easy to show that

(3.5)'' $e^{B\Delta t/2}e^{A\Delta t}e^{B\Delta t/2}$

approximates (3.4) with an error of order $(\Delta t)^3$. The nth power of the operator (3.5)'' telescopes neatly as follows:

$$(e^{B\Delta t/2}e^{A\Delta t}e^{B\Delta t/2})^n = e^{-B\Delta t/2}(e^{B\Delta t}e^{A\Delta t})^n e^{B\Delta t/2}.$$

This suggests to use as approximation to $e^{(A+B)t}$:

$$(3.6)'' \qquad\qquad S_{-B/2}(S_B S_A)^n S_{B/2}.$$

Clearly, $(3.6)''$ is only insignificantly more costly to evaluate than (3.6) raised to the nth power.

The fractional step method has the great advantage of breaking up a complicated problem into simpler parts. Very often some of the components are so simple that the initial value problem associated with one fraction can be solved explicitly in terms of trigonometric functions. This possibility has been exploited by Fred Tappert (see [7]) who combined it with an efficient way of evaluating explicit solutions: Suppose the operator A appearing as one of the fractions is a differential operator in x, with constant coefficients, and suppose that u is periodic in x. Then the fractional step e^{At} can be evaluated by using Fourier series, i.e., by introducing the eigenfunctions of A. Writing

$$(3.7) \qquad\qquad e^{At}u(x) = \sum_p a_p(t)e^{ipx},$$

we have

$$(3.8) \qquad\qquad a_p(t) = e^{\lambda_p t}a_p(0),$$

where

$$(3.9) \qquad\qquad \lambda_p = A(ip)$$

and

$$(3.10) \qquad\qquad a_p(0) = \int u(y,0)e^{-ipy}\,dy.$$

Substituting (3.10) and (3.8) into (3.7) we get

$$(3.11) \qquad\qquad (e^{At}u)(x) = \int K(x-y,t)u(y,0)\,dy$$

where

$$(3.12) \qquad\qquad K(z,t) = \sum e^{A(ip)t+ipz}.$$

To turn (3.11) into a numerical scheme we replace the integral by an approximate sum and set $t = \Delta t$:

$$(3.13) \qquad\qquad (S_A u)_j = \sum_k K_{j-k}u_k$$

where u_k denotes the value of $u(x,0)$ at the kth meshpoint, and $K_l = \Delta x K(l\Delta x, \Delta t)$. Again, we must ask the question: How much computing does it take to evaluate S_A using (3.13)?

Denote the number of space points by m; clearly, for each value of j, evaluating (3.13) takes m multiplications and additions. Since this has to be done for m values of j, the total number of additions and multiplications is m^2. This is in contrast to explicit or implicit schemes where the total number of operations per time step is of the order m. Tappert has observed that the number of operations can be reduced if instead of using formula (3.11) one approximates the integral (3.10) for $a_p(0)$ by a sum:

$$(3.14) \qquad a_p(0) \simeq \Delta x \sum_k e^{ipk\Delta x} u_k.$$

Substitute this approximation into (3.8), and then substitute (3.8) into (3.7), restricting the summation to $|p| \leqslant m$. Again we must count the number of operations: To evaluate formula (3.14) for any value of p we need m multiplications and additions; since this has to be done for $2m + 1$ values of p, the number of operations needed to evaluate all the a_p appears to be proportional to m^2. Similarly, to evaluate the sum (3.7), $|p| \leqslant m$, at all m meshpoints $x = k\Delta x$ appears to take another $2m^2$ multiplications and additions. So, apparently, nothing has been gained, except avoiding the evaluation of the function $K(z, \Delta t)$ (which has to be done for only m distinct values of z). However, Cooley and Tukey (see [1]) have shown that the Fourier transforms (3.7) and (3.14) can be evaluated using not $2m^2$, but only $2m \log m$ operations! This is a key element in Tappert's method.

As the last example, we present an elegant method due to Varga (see [9]) where an initial value problem is solved not by applying n times an operator S but at one stroke.[2] The method is applicable to equations of the form $u_t = Au$, where A is a symmetric, negative operator independent of t; parabolic equations are par excellence of this form. The idea of the method is this: Since the exact solution operator is e^{At}, an approximate operator would be chosen of the form $r(At)$ where $r(s)$ is a rational approximation to the exponential function e^s. Let $E(\lambda)$ be the spectral resolution of A:

$$A = \int \lambda dE(\lambda).$$

By the functional calculus,

$$e^{At} = \int e^{\lambda t} dE(\lambda), \qquad r(At) = \int r(\lambda t) dE(\lambda),$$

so that the difference of exact and approximate solutions is

---

[2]To paraphrase Armstrong, many small steps for mankind, a giant leap for Varga.

$$\int [e^{\lambda t} - r(\lambda t)] \, dE(\lambda).$$

Since A was assumed negative, its spectrum lies on $\lambda < 0$; thus we get the following error estimate:

(3.15) $$\|e^{At} - r(At)\| \le M$$

where $M = \max_{s \le 0} |e^s - r(s)|$. Clearly, the better $r(s)$ approximates e^{-s}, in the maximum norm on the negative axis, the sharper the error estimate (3.15) is. Therefore [9] chooses for $r(s)$ the best Chebychev approximation to e^{-s} on $s \le 0$. There are effective algorithms for computing these best approximations. How expensive is it to evaluate $r(A)u$, where $r(s) = p(s)/q(s)$ is a rational function and A a symmetric matrix of high order? The straightforward way of doing it is to evaluate $v = p(A)u$, by, say, Horner's method, then compute $q(A)$, and compute $q(A)^{-1}v$. In the kind of applications we have been considering A is a difference approximation to a differential operator, and therefore A is *sparse*; i.e., most of its entries are zero. In forming $q(A)$ we lose this sparseness, which makes the inversion of $q(A)$ more expensive. Therefore [9] advocates the following algorithm:

Factor q into linear factors:

$$q(s) = \prod_{1}^{d}(s - s_j),$$

d the degree of q. Evaluate $q(A)^{-1}v$ as

(3.16) $$q(A)^{-1}v = \prod(A - s_j)^{-1}v.$$

In this method a single inversion of a fat matrix is replaced by d inversions of sparse matrices; this is economical if the order of the matrix A is large.

Notice that the operations in (3.16) are nested and have to be performed serially. If a parallel computer is available, it might be more advantageous to make use of the partial fraction expansion of $r(s)$.

REFERENCES

1. J. W. Cooley and J. W. Tukey, *An algorithm for the machine calculation of complex Fourier series*, Math. Comp. 19 (1965), 297–301. MR 31 #2843.

2. P. Halmos, *A Hilbert space problem book*, Van Nostrand, Princeton, N. J., 1967. MR 34 #8178.

3. P. D. Lax and B. Wendroff, *Difference schemes for hyperbolic equations with high order of accuracy*, Comm. Pure Appl. Math. 17 (1964), 381–398. MR 30 #722.

4. D. W. Peaceman and H. H. Rachford, Jr., *The numerical solution of parabolic and elliptic differential equations*, J. Soc. Indust. Appl. Math. 3 (1955), 28–41. MR 17, 196.

PETER D. LAX

5. R. D. Richtmyer and K. W. Morton, *Difference methods for initial-value problems*, Interscience, New York, 1967. MR **36** #3515.

6. D. J. Rose and R. A. Willoughby, *Sparse matrices and their applications*, Plenum Press, New York, 1972.

7. F. Tappert, *Numerical solution of the KdV equation and its generalizations by the split-step Fourier method*, Lectures in Applied Math., vol. 15, Amer. Math. Soc., Providence, R. I., 1974, pp. 215–216.

8. N. N. Yanenko, *The method of fractional steps*, Springer, New York, 1971.

9. W. J. Cody, G. Meinardus and R. S. Varga, *Chebychev rational approximations to e^{-x} in $[0, +\infty]$ and applications to the heat-conduction problems*, J. Approximation Theory 2 (1969), 50–65. MR **39** #6536.

COURANT INSTITUTE OF MATHEMATICAL SCIENCES

Proceedings of Symposia in Applied Mathematics
Volume 20
1974

THE UNEXPECTED IMPACT OF COMPUTERS
ON SCIENCE AND MATHEMATICS

BY

THOMAS E. CHEATHAM, JR.

In the past two decades, digital computers have had a very dramatic impact on the teaching, research, and practice of the sciences and mathematics. At any point in time, as computers were becoming more and more accessible, as well as becoming larger and cheaper, a certain aspect of their impact has been clear—the scientist could compute more at less cost with better facilities. In this paper, however, I want to discuss a different kind of impact, an impact which in many cases goes to the very core of a particular science and whose result has at times been a new perspective and at others has been a whole new theory. Before doing this, however, let me discuss briefly the traditional reasons for computing and the role of scientific computation.

Why compute? The traditional reasons for using digital computers at all, in the sciences or in general, have been those of cost and of necessity. That is, to handle a large payroll, to manage the federal tax records, to compute approximate solutions to complex differential equations, to generate various statistical measures concerning large volumes of data, to invert large matrices, and so on, it is not cost-effective to employ people and hand calculators. The use of high speed digital computers is simply mandatory if anyone is to pay the bill for these kinds of computations.

In recent years, as computers have become sufficiently large, cheap, and reliable, we can cite a number of situations in which computation—employing high speed computers in some system—was a must. Here we have in mind such applications as missile tracking, the moon landing, many production control facilities, and so on. Without modern computers, these projects simply could

---

AMS (MOS) subject classifications (1970). Primary 94–04.

not have been undertaken. This calls to mind the story circulated several years ago concerning the Bell System automated exchanges. Somewhere along about a decade ago the telephone traffic in this country reached a level where, if all connections had been handled by operators, it would have required the services of half the people over the age of sixteen in this country to keep the system going.

The traditional role in scientific computation. As Oettinger noted a few years ago [3]:

> A physical theory in the language of mathematics often becomes dynamic when written as a computer program; we can explore its inner structure, confront it with the experimental data and interpret its implications much more easily than when it is in the static form. In disciplines where mathematics is not the prevailing mode of expression the language of computer programs serves increasingly as the language of science.

He continues:

> The advance of science has been marked by a progressive and rapidly accelerating separation of observable phenomena from both common sensory experience and theoretically supported intuition. Anyone can make at least a quantitative comparison of the forces required to break a matchstick and a steel bar. Comparing the force needed to ionize a hydrogen atom with the force that binds the hydrogen nucleus together is much more indirect, because the chain from phenomenon to observation to interpretation is much longer. It is by restoring the immediacy of sensory experience and by sharpening intuition that the computer is reshaping experimental analysis.

This trend of which Oettinger spoke has, if anything, accelerated over the past several years. As computers have become more easily available more scientists have taken advantage of them. Their use by the psychological and social sciences, for example, has accelerated dramatically, and there is voiced more and more concern about their impact on society [5]. It is also worth noting that the development of small, inexpensive "mini-computers" and the possibility of employing some fraction of a large computer through time-sharing has resulted in more and more "closed loop" systems by incorporating computers directly in various experiments using dynamic feedback to control the future progress of the experiment.

In the straightforward use of computers in the various sciences, the situation is either that we have some physical phenomena described in mathematical equations and we construct a computer program to solve these equations, or that we construct a computer program to directly model the physical situation. In any event, the "results" we want are deeper understanding of the phenomena, often in the form of various numerical results produced by variation of certain "interesting" parameters.

Models which have been used to describe these phenomena have been drawn from various areas of mathematics; such traditional mathematical problems as the solution of differential and integral equations, inversion of matrices, determination of zeros, maxima or minima of expressions, solution of systems of inequalities, and so on, have been required. Interest in these problems had, and has continued to have, a considerable impact on mathematics—particularly numerical mathematics. For years numerical analysts had concentrated their attention on techniques now considered as quaint and arcane because they were expected to be carried out by people with hand calculators. Years ago efforts were refocused on iterative techniques, concern with automatic error control, and so on. But this kind of impact on mathematics was hardly "unexpected"—it was clear two decades ago that the whole direction of numerical mathematics and of other supportive areas of mathematics would change.

The size and power of digital computers has grown (and, concomitantly, the cost has decreased) by orders of magnitude over the past two decades, thus permitting more and more bigger and bigger scientific computation to be performed. By now, the capacity of computers is far, far beyond what humans can do. But, as the power and ability to compute has grown, so has the appetite of scientists for computation. We seem always to be possessed of problems whose solution is beyond the power of current hardware, software and numerical techniques: Just as we can finally handle, say, reactor calculations more or less adequately, we note that global weather forecasting is well beyond our current facilities.

The role of a theory. To employ a computer we *require* a theory. It is not necessary that this be in the form of a physical theory which in turn can be captured via certain mathematical equations which in turn can be captured via appropriate numerical algorithms which provide approximate results. Looked at another way, to compute at all we must have a program which directs a computer in a completely unambiguous way and often in such detail as to overwhelm the user. That is, a computer program is simply an executable version or representation of some algorithm. It is often the case that the computer program version of an algorithm is not "fit for human consumption" and that a version of the algorithm which deals in data objects and operations much closer to the scientist's area of discourse is the preferred form. Indeed much of the work in programming languages and their processors over the past several years has had as its goal that of providing a language close to that natural to a particular area of science while at the same time being susceptible of mechanical translation into an efficient computer program. But while having the appropriate level of description is important, our point here is that whatever level is provided, the crucial point is that the algorithm must describe in a precise and unequivocal way *exactly* what process is to be performed on *exactly* what data in order to achieve the desired result. It is this exactness, this precision of algorithmic specification required to permit computation at all which we feel has had the kind of impact which was, at least by most scientists, unexpected. And, many scientists have found that describing a procedure to be carried out by some assistant in a laboratory is quite an entirely different matter from describing a procedure appropriate for a digital computer. The amount of ambiguity and lack of detail which can be accomodated by intelligent human beings is very great indeed; that which can be accomodated by a computer is zero.

This is the heart of our thesis—that the requirement for exact algorithms to model some aspect of the world of interest to a scientist has had an unexpected, and sometimes profound, impact on that scientist and his understanding of the world. Let us consider some examples.

Organic molecular synthesis. It is the business of large numbers of organic chemists to develop methods for synthesizing some particular molecule of interest. Put simply, they begin with some molecule whose production is desired and end with a procedure—a synthesis technique—which will produce appropriate quantities of the desired molecule in one or more chemical processes from material readily at hand. Indeed, large numbers of synthesis techniques have been developed—for medicines like insulin, for various industrial products like plastics and so on—and the needs for such syntheses continue.

Several years ago Professor E. J. Corey at Harvard asked whether or not some of the detail of developing synthesis techniques could be done by a computer. Corey imagined a man/machine system in which a chemist would present to a computer some molecule to be synthesized, whereupon the computer and chemist would work together to develop a synthesis technique. It is hoped that as time progressed, more and more chemical knowledge would be encoded in the computer program and less and less help would be required of the chemist.

Corey and his colleagues have spent several years developing such a computer program and have developed a man/machine system now operational for practical organic molecular synthesis. At present, the "machine" part is quite good and capable of generating simple syntheses. With the chemist also working, the system can generate some quite sophisticated syntheses, and as more chemical knowledge is encoded in the "machine" part it gets better and better, needing to depend less and less on the chemist.

But Corey's success is not our point. The point is that in attempting to develop a computer program to do organic synthesis, Corey and his group found that the traditional view of organic chemistry was simply not adequate. Their theory, if you will, was not appropriate and what was in essence a new theory was required. At this time their new view of organic chemistry is having very important impacts on chemistry—the kinds of questions that are asked in research in organic chemistry and the methods used for teaching organic chemistry are different—an impact which was not at all suspected when their project commenced.

Symbolic mathematics. There has been interest since the very early days of large scale computers in devising programs to do certain symbolic calculations—including such tedious tasks as symbolic differentiation and integration of functions, taking limits, simplifying expressions, and so on. Of course the problem of symbolic differentiation is very simple and programs have existed for years to do that job. However the other problems have proved rather more challenging and it is only very recently with the MACSYMA system [1] that a general facility for doing nontrivial symbolic mathematics has been available.

The first attempts to do such things as symbolic integration were, in a very strong sense, trying to copy what people do when they are confronted with some expressions to integrate. The style of specification was very much that now referred to as pattern-replacement and basically involved separating out a large number of special cases. So long as this basic approach was employed, no program for symbolic integration was anything more than a curiosity. There were interesting side effects of work on this problem, as there were from working on many other problems of the sort often characterized as "artificial intelligence" problems. These had to do with the development of programming languages and systems rather better suited to such problems than the traditional ALGOL, FORTRAN or COBOL facilities had been and with the development of interesting pattern/replacement facilities, techniques for developing efficient search strategies, and so on.

However, real progress in providing a practical program for symbolic integration basically awaited the results of Risch [4] who developed a finite algorithm for the integration of an arbitrary expression from a certain class of expressions. An interesting aside here is the kind of mathematics from which Risch developed his eminently practical result, namely algebraic geometry and differential algebra, two areas of mathematics which have seldom been known for their practicality.

The relevant point here is that again the early attempts to mechanize integration showed a basic lack of a theory of the proper sort (there was, of course, a rich theory of "integration" but that theory was not of much help in actually *doing* a given integration). Risch's theory has already had an impact in permitting practical symbolic computations and there are some who feel that it may also have a very profound impact on the way mathematics is taught. Indeed, there has already been some discussion of using Risch's algorithm in an introductory calculus course rather than teaching the traditional "cut and try" or "table look-up" methods for integration.

Special functions. During the nineteenth century the development of mathematics in Britain was mostly concerned with the detailed study of the properties of classes of special functions defined by ordinary differential equations found useful in certain physical models. In time, the information about these special functions became encyclopedic in scope. However, there were some basic difficulties, as for example no reasonable set of basic functions is closed under integration, and one cannot usually solve a transcendental equation in terms of a fixed set of functions. Furthermore, new physical problems are often encountered which do not possess models reducible to those involving functions whose properties are already known. Thus, one is always in the business of studying the properties of more and more special functions.

Using machinery developed by pure mathematicians over the past several decades, it may be possible to achieve the goal of the British school not by studying properties of particular classes of functions, but by developing algorithms for generating desired properties directly from their defining differential equations [2]. We note that this kind of approach would probably not have been conceived without the past decade of work on programs for symbolic mathematics and the "algorithmic" attitude and style of specification induced by computers.

Medical diagnosis. A number of computer programs which take data from a patient and produce a medical diagnosis have been developed recently in such areas as diabetes, stroke analysis, determining rare salt balance, and so on. The situation here has been similar to that encountered by the organic chemists, namely that a theory adequate to developing the desired computer programs was lacking and the construction of the required theory (the precise diagnosis programs) gave specific insights into the medical field which have had an important and unexpected impact.

Communication of complex algorithms. Developments such as programs for medical diagnosis and organic molecular synthesis have in turn had an impact on computer science which was unexpected, and that is the need to communicate to certain specialists (for example, medical doctors) the details of some algorithms basically represented as a computer program, but to do so in a form or language which they can understand and evaluate. The problem of "documenting" computer programs has been a persistent problem to data processing installation managers for years, but the problems posed by such applications as medical diagnosis and organic synthesis are rather more difficult.

It is always possible to find a few individuals who are expert in some specialized area and know enough about programming to write programs such as those for medical diagnosis and organic molecular synthesis mentioned above. However, it is seldom the case that these programs or systems remain static; they get more and more complex as more and more knowledge is added to them. Furthermore, at some point it is usually necessary to incorporate the specialized knowledge of experts who are not trained in computer programming. For these people, determining exactly what has been done and exactly what decision procedures have been used is very difficult because they cannot readily understand the computer program. The same problem arises for anyone who wishes to judge the adequacy or correctness of some procedure proposed by a colleague or who is going to use the programs for teaching medical diagnosis. He must be able to *read* the program and to understand it, but surely without having to understand irrelevant aspects of computers and computer programming. Thus, the problem posed to computer scientists is that of developing representations of some algorithm—a computer program—suitable for consumption by medical doctors (or organic chemists, or what have you) and of developing facilities which would permit, say, the medical doctor to explore and understand very complex algorithms. In a sense this requirement presents an unexpected twist to computer scientists as their concern for some time has primarily been with providing languages in which programs are to be *written*, not those in which they are to be read.

Some expected impacts which did not work out. The application of computers in the various sciences has had its failures as well as its successes, and there are several areas where successful solution of some problem by a computer was "just around the corner" but which has by now been given up or at least admitted to be much harder than initially anticipated. These applications include such things as mechanical translation of natural languages, recognition of handwritten text, construction of sophisticated robots, and even playing chess at the level of the best human players (in real time).

In all these cases the initial optimism of producing a practical system turned to pessimism as it was discovered that we simply did not possess an *adequate* theory—and, as a corollary, that human beings are very remarkable indeed in being able to solve any number of very sophisticated problems *without* a theory, or at least a theory which can be set down. There is probably no doubt that these and other difficult problems will eventually be "solved" by computers and that the development of an adequate theory as an algorithm susceptible of computer implementation is what is required.

Some conclusions. Let us review our basic thesis and then propose some conclusions which can be drawn. Our basic thesis is that in many areas the business of developing a theory or of adapting an extant theory as an algorithm suitable for computer implementation has had the unexpected impact of providing a new perspective in the subject and that in many cases this has had many important ramifications. We also note that in this sense the computer—the device—is entirely irrelevant; that is the algorithmic style of thinking which is important.

One obvious conclusion from this is that this should be understood by those scientists involved in developing a new theory—that if the form of the theory is suitable for more-or-less straightforward computer implementation then they are probably much better off than they are with a theory which is difficult to so implement.

Perhaps a less obvious and more controversial conclusion has to do with the educational process. That is, if the algorithmic style of specification or theory is indeed important, then how do we transmit this to a broad audience and train our future scientists as well as our future medical doctors, lawyers, businessmen and so on who might be expected to devise and to understand an algorithmic style of specification. Traditionally, it is through mathematics courses (for example, plane geometry) that the student is introduced to theory—axiom systems, logical deductions, and so on. Perhaps we should explore the possibility of supplementing or even replacing some traditional science and mathematics courses with courses more orientated to the use of algorithms. One of the problems with the traditional mathematical approach to modelling is that it often requires sophisticated mathematics to deal with even the simplest physical models.

By employing the algorithmic approach, however, we are able to obtain meaningful and precise models. The resulting models often have the advantage of clarity as well, surely a desirable goal.

BIBLIOGRAPHY

1. Mathlab Group, MACSYMA *reference manual*, Version 5, Project Mac, MIT, June 1973.

2. Joel Moses, *Towards a general theory of special functions*, Comm. ACM **15** (1972), 550–554.

3. Anthony G. Oettinger, *The uses of computers in science*, Scientific American **215** (1966), 161–172.

4. Robert H. Risch, *The solution of the problem of integration in finite terms*, Bull. Amer. Math. Soc. **76** (1970), 605–608.   MR **42** #4530.

5. J. Weizenbaum, *On the impact of the computer on society*, Science **176** (1972), 609–614.

HARVARD UNIVERSITY

CONTRIBUTED PAPERS

Proceedings of Symposia in Applied Mathematics
Volume 20
1974

COMPUTATIONAL COMPLEX ANALYSIS

BY

PETER HENRICI

The essence of my paper will consist in outlining an attempt to rejuvenate the traditional complex analysis course by transplanting some branches of this fertile mathematical tree into the vast new industrial park of computation. As a conscientious teacher I would like to begin by providing the proper motivation for my work.

On the surface, mathematics today seems in excellent shape. Never before in our history have so many theorems been proved per unit time, nor has the general technical level of these theorems been so high.

Yet it is also commonplace that all is not well with mathematics. Some of the people on whom we depend for a livelihood--undergraduate students, prospective Ph.D's, prospective customers and, worst of all, prospective employers—turn their backs on us.

I am just a small country mathematician—by which I mean a mathematician from a small country, and not even quite a mathematician at that—and thus I should approach this subject only with extreme prudence, or best not at all. If I nevertheless decided to stick my neck out, I did so because I thought that my distant vantage point would offer some advantages. As it turns out, what I have to say is not so very different from what the other papers in this volume have to say.

In my view, the present crisis in mathematics is, at least in part, due to the fact that an equilibrium between two opposing forces has been disturbed. Mathematics owes much of its vitality to several pairs of opposing forces, or polarities, all of which have proved fruitful in the past. Suffice it to mention the polarities between

---

AMS (MOS) subject classifications (1970). Primary 90A30.

pure	versus	applied
abstract	versus	concrete
theory-oriented	versus	problem-oriented

mathematics. Somebody in another paper talks about static versus dynamic mathematics, which may be a somewhat loaded formulation of a similar polarity. The polarity which I have in mind and whose equilibrium is disturbed is positively correlated to all of the above, yet different. After some consulting of dictionaries I decided to call it the polarity between *dialectic* and *algorithmic* mathematics.

Dialectic mathematics is a rigorously logical science, where statements are either true or false, and where objects with specified properties either do or do not exist. *Algorithmic mathematics* is a tool for solving problems. Here we are concerned not only with the existence of a mathematical object, but also with the credentials of its existence. *Dialectic* mathematics is an intellectual game played according to rules about which there is a high degree of consensus. The rules of the game of *algorithmic* mathematics may vary according to the urgency of the problem on hand. We never could have put a man on the moon if we had insisted that the trajectories should be computed with dialectic rigor. The rules may also vary according to the computing equipment available, as we have learned from Traub [10]. *Dialectic* mathematics invites contemplation. *Algorithmic* mathematics invites action. *Dialectic* mathematics generates insight. *Algorithmic* mathematics generates results.

As an example, take the fundamental theorem of algebra. A dialectic mathematician probably will be happy with one of the many nonconstructive proofs, say via Liouville's theorem. The algorithmic mathematician would prefer a method that is guaranteed to produce the zeros to any desired accuracy.

As another illustration, consider applied mathematics, which I would like to define as the *study of models*. While some dialectic mathematics may be required to set up the model, algorithmic mathematics is required to work with the model. If the model consists of a boundary value problem for, say, a partial differential equation, the dialectic mathematician would probably be satisfied by demonstrating existence and uniqueness of the solution. However, to the algorithmic mathematician it matters a great deal whether the solutions exist (a) merely because assuming the contrary would lead to a contradiction, (b) by virtue of the axiom of choice, or (c) because there is a computer program to construct it. If there is a program, it matters whether there is an a priori time limit for its execution. If there is a time limit, it matters whether it is 10,000 years or a few seconds, as is the case for some recent programs for solving Poisson's equation in the unit square using 2^{12} lattice points [1].

That mathematics has, and should have, algorithmic content seems to have been taken for granted, say, at the time of my compatriot Leonhard Euler. In the intervening two centuries our science seems to have lost some of its algorithmic flavor. Until 1950 this could be explained, and even justified, by the lack of new tools for carrying out algorithms. Today this excuse no longer holds. My main thesis is that *the traditional curriculum as offered today is deficient in algorithmic content. There should be an increase in the teaching of algorithmically oriented mathematics.*

Dialectic mathematicians sometimes take the attitude that all is well if the students are taught good pure mathematics [7]. Any student thus equipped, so this theory goes, would then find it easy to learn all the applications and all the algorithms he ever wanted to know. To this I would reply that, in all fairness, the algorithmic talent is not merely a subtalent of the dialectic talent. There are very specific examples of attempts by outstanding dialectic mathematicians to solve computational problems that later were solved in a much more satisfactory manner by algorithmically oriented researchers [9]. As a well-known example, consider the algebraic eigenvalue problem. The classical dialectic approach is via the characteristic polynomial. From the point of view of actual computational practice, this approach has long been disregarded as utterly impractical [13]. Numerical analysis today offers several feasible alternatives—yet a student of such dialectically excellent textbooks on linear algebra as those by Hoffmann and Kunze [5] or Greub [3] is still left with the impression that the eigenvalue problem (a vital problem in many applications of mathematics) is solved by first constructing the characteristic polynomial. Mathematics certainly contributed to our success in space, but this success is due less to our dialecticians than to humble programmers who would never dream of being invited to a meeting of the American Mathematical Society. I believe that in the teaching of mathematics the algorithmic talent should be cultivated independently of and in addition to the algorithmic talent.

By "cultivating the algorithmic talent" I do not necessarily mean courses in recursive function theory, or even in the theory of algorithms. Much of this work is simply an extension of the dialectic method to the area of algorithms, and the results are frequently of a negative nature. A colleague of mine in a very famous paper proved that in recursive analysis one cannot multiply certain recursive decimal fractions by the positive integer 3 [8].

On the other hand, I am not necessarily advocating an increase in numerical analysis courses. A course, say, on numerical methods in differential equations will automatically take some pressure off the theoretical differential equations course, and thus decrease its algorithmic content. Somewhat reluctantly, I should also point out that some parts of numerical analysis have become quite dialectic.

Rather, I would like to see much of the content of the numerical analysis courses as now taught to become part of the corresponding theoretical courses. At the ETH Zürich we have for some six years now been teaching a basic three-semester course in engineering mathematics that also covers elementary numerical methods, including the solution of two-dimensional boundary value problems. (By now even those instructors whose training was mainly in dialectic mathematics know this material.) The same should be done for students of mathematics, and also on a higher level.

Complex analysis is a point in case. From its very inception, the theory of analytic functions had a significant algorithmic component. According to the article by Osgood in the *Enzyklopädie der mathematischen Wissenchaften* much of the motivation for the early development of complex analysis came from a desire to understand the so-called Lagrange-Bürmann series, which Lagrange devised to solve Kepler's equation

$$E - M = \epsilon \sin E.$$

One of the early successes of complex function theory was the clarification of the convergence behavior of the *Taylor series,* a result of tremendous algorithmic potential. Cauchy's theory of residues likewise has proved to have an enormous algorithmic impact. A good deal of the theory of elliptic functions had its motivation in the problem of finding rapidly converging expansions for elliptic integrals. One of Riemann's contributions to complex analysis was to introduce a unifying point of view into the maze of formulas representing solutions of the hypergeometric differential equation. The theory of special functions, pushed to a high degree of perfection in the last century, provided algorithmic solutions to innumerable problems of mathematical physics.

All of this is old hat, of course. Some of the topics mentioned are still taught. But do we really wish to restrict our teaching of algorithms to those that were invented 150 years ago? In a book [4] on which I have been working since 1961 and whose first volume is now in production, I have tried to call attention to some algorithmically potent areas of complex analysis that are of more recent origin.

In my book, complex analysis is based on the *theory of power series.* Of the two classical approaches to function theory, the "analytic" approach of Weierstrass is definitely closer to computation than the "geometric" approach of

Riemann. As has been pointed out repeatedly, a computing machine cannot really deal with functions defined on a continuum, but it can very readily handle sequences of power series coefficients and perform all kinds of algebraic operations on them. In my treatment of power series I make a sharp distinction between the algebraic and the analytic aspects of the theory. This should teach the student the increase in depth when proceeding from the former to the latter. But already the formal theory offers much that is of computational interest, especially if we consider nontrivial operations such as raising a power series to an arbitrary power, or the inversion of a power series. Concerning the problem of inversion, it does not seem to be generally known that the classical Lagrange-Bürmann expansion already holds in the formal context, even in the case of several variables. This formula implies a ready-made algorithm for solving the inversion problem that appears to be much simpler than the traditional method of undetermined coefficients. In the convergent case the Lagrange-Bürmann series can be applied to the solution of all kinds of transcendental equations, such as $\tan x = \mu x$. Another application is the determination of the zeros of a polynomial that is perturbed by another polynomial. Many of these applications are classical, but because of the amount of computation needed it is unlikely that they have ever been implemented on a numerically nontrivial scale.

I have mentioned the *theory of residues* as having algorithmic appeal, but it may be hard to see how the computer can contribute to the perfection of Cauchy's residue formula. Consider, however, the problem of summing "one-sided" infinite series. Applying residue methods, we obtain the Plana summation formula

$$\sum_{n=0}^{\infty} f(n) = \tfrac{1}{2}f(0) + \int_0^\infty f(x)\,dx + i \int_0^\infty \frac{f(iy) - f(-iy)}{e^{2\pi y} - 1}\,dy$$

expressing the sum as an integral plus a correction term. (To approximate the discrete by the continuous is a typical feature of algorithmic analysis.) The first integral can usually be done analytically, while the second must be computed numerically. (A student in Zürich seemed to have fun using this formula to hunt for some nontrivial zeros of the Riemann zeta function.)

In the theory of *conformal mapping,* the basic algorithmic problem is the construction of the mapping function for a given simply or doubly connected region. There are many approaches; see the excellent book by Gaier [2]. All of them require a certain sophistication; hence, this topic is perhaps best dealt with in a graduate seminar. I could think of a seminar in which one and the same problem would be solved by several students, each student using a different method, and notes would be compared. On a much more elementary level in conformal

mapping, *Moebius transformations* (or bilinear transformations) offer themselves
for algorithmic exploitation. They enjoy the pleasant property of mapping circular
regions onto circular regions, which fact provides the basis of what I call *circular
arithmetic*. This is an extension to the complex plane of *interval arithmetic,* a
device for error control in digital computations. Circular arithmetic has applications in the classical geometry of polynomials as well as in some modern algorithms
for zero-finding; moreover, it is a useful device in the theory of continued fractions.

 Continued fractions are another rich source of algorithmic mathematics. Far
from being a mere tool in constructive number theory, they have important applications in the approximation of functions, in the summing of asymptotic expansions, and in stability theory. Continued fraction representations of certain
elementary and higher transcendental functions sometimes converge in large portions of the complex plane where the usual power series expansion fails.

 Turning back to elementary matters in complex analysis, even a fact as basic as the *principle of the maximum* has algorithmic implications, for it essentially
guarantees that any method for solving $f(z) = 0$ that is based on minimizing
$|f(z)|$ must be successful. As a result, the theory of such "methods of descent"
is much neater for complex functions than for real functions. This theory forms
the basis for a polynomial solving algorithm inspired by Rutishauser and implemented by Kellenberger, which may well be one of the fastest polynomial solving
routines in existence.

 Winding numbers are another fundamental concept of complex analysis. It
is frequently pointed out that winding numbers, via the principle of the argument,
can be used to determine zeros of functions. The implementation is not quite as
simple as it first appears, because it is not clear a priori how accurately the winding
number integral must be computed. It should interest the student to know how
Hermann Weyl overcame this difficulty in his constructive proof of the fundamental
theorem of algebra [12]. Another algorithmic application of winding numbers is
the Schur-Cohn algorithm for determining the number of zeros of a polynomial in
a disk, which forms the basic ingredient of another successful polynomial solver
due to Lehmer [6].

 One of the favorite topics in the traditional numerical analysis course is the
treatment of *iteration* as a tool for finding *fixed points*. A major role is played by
the Lipschitz condition (with $L < 1$) which is to be satisfied by the mapping.
This is required both for the uniqueness of the fixed point, and for the convergence
of the iteration sequence. It is seldom pointed out that the situation is much
nicer in complex function theory. All we need is the following: *Let f be
analytic in a simply connected domain S and continuous in the closure S' of S,*

and let $f(S')$ *be a bounded set contained in* S. *Then* f *has a unique fixed point in* S, *and the iteration sequence converges to the fixed point for every choice of the starting value in* S'. The proof requires the Schwarz lemma and the Riemann mapping theorem. Even more general versions of the above lemma are true, but already the above version permits a full treatment of most current iteration functions.

As a last topic of algorithmic appeal, I call attention to the *connection between the Taylor coefficients* of an analytic function and its *singularities*. This was already used by Daniel Bernoulli, another compatriot from Basel, in what must be one of the oldest methods for solving polynomial equations. The quotient-difference algorithm due to Rutishauser is based on the same idea. This algorithm can be used, for instance, to approximate simultaneously all zeros of transcendental functions such as the Bessel functions. By methods which were accessible to Euler, it thus solves a problem which was considered by Euler but not solved with the same perfection (cf. [11, p. 500]). It surely can do no harm to a student of complex analysis to become acquainted with a modern algorithm of such classic simplicity.

All of the above topics invite active participation by the student in the form of experimental computation. Here I wish to point out that dialectic mathematics today has been pushed to such a high degree of perfection that it often leaves a student who is not exceptionally gifted or creative with a sense of utter frustration. In algorithmic analysis even a student who is not outstanding can make a real contribution by comparing various implementations of an algorithm and testing their efficiency and stability.

In this paper I have concentrated on complex analysis, in fact on rather elementary complex analysis. But this is just one of many mathematical disciplines where the algorithmic content should be strengthened. An even greater problem child is linear algebra, where the gulf between dialectic and algorithmic content is especially wide. Another sad case is number theory. I am afraid that, in today's mathematical education, number theory often merely serves as a source for illustrations of concepts of abstract algebra. I feel that it is deplorable to neglect the algorithmic aspects of the subject, such as tests for primality, diophantine equations, Pell's equation, etc. Even at the high school level it is possible to treat number theoretical questions with algorithmic content, such as the representations of numbers by periodic or nonperiodic decimal fractions, or the simple tests for divisibility. Judging from the comments of experienced teachers, the algorithmic solution of problems in number theory offers intellectual pleasure even to those who are not equipped to understand the underlying abstractions.

In conclusion, let me express the essence of this paper in a well-known quotation from Thomas H. Huxley:

"The great end of life is not knowledge, but action."

BIBLIOGRAPHY

1. B. L. Buzbee, G. H. Golub and C. W. Neilson, *On direct methods for solving Poisson's equations,* SIAM J. Numer. Anal. **7** (1970), 627–656.   MR **44** #4920.

2. D. Gaier, *Konstruktive Methoden der konformen Abbildung,* Springer Tracts in Natural Philospohy, vol. 3, Springer-Verlag, Berlin, 1964.

3. W. H. Greub, *Linear algebra,* 2nd ed., Die Grundlehren der math. Wissenschaften, Band 97, Academic Press, New York; Springer-Verlag, Berlin, 1963.   MR **28** #1201.

4. P. Henrici, *Applied and computational complex analysis.* Vol. I: *Power series, integration, conformal mapping, location of zeros,* Wiley, New York, 1974.

5. K. Hoffman and R. Kunze, *Linear algebra,* Prentice-Hall, Englewood Cliffs, N. J., 1961.

6. D. H. Lehmer, *A machine method for solving polynomial equations,* J. Assoc. Comput. Mach. **8** (1961), 151–162.

7. B. Scarpellini, *Statement made to Planning Committee for Mathematics and Astronomy,* University of Basel, February 17, 1972.

8. E. Specker, *Nicht konstruktiv beweisbare Sätze der Analysis,* J. Symbolic Logic **14** (1949), 145–158.   MR **11,** 151.

9. A. Talbot, *The evaluation of integrals of products of linear system responses.* I, II Quart. J. Mech. Appl. Math. **12** (1959), 488–503, 504–520.   MR **22** #9816a, b.

10. J. F. Traub, *An introduction to some current research in numerical computational complexity,* Proc. Sympos. Appl. Math., vol. 20, Amer. Math. Soc., Providence, R. I., 1974.

11. G. N. Watson, *A treatise on the theory of Bessel functions,* 2nd ed., Cambridge Univ. Press, Cambridge, 1944; Macmillan, New York, 1944.   MR **6,** 64.

12. H. Weyl, *Randbemerkungen zu Hauptproblemen der Mathematik.* II. *Fundamentalsatz der Algebra und Grundlagen der Mathematik* Ges. Werke **2** (1968), 444–452.

13. J. H. Wilkinson, *Modern error analysis,* SIAM Rev. **13** (1971), 548–568.   MR **46** #4708.

EIDGENÖSSISCHE TECHNISCHE HOCHSCHULE ZÜRICH

Proceedings of Symposia in Applied Mathematics
Volume 20
1974

COMBINATORIAL GAMES WITH AN ANNIHILATION RULE

BY

AVIEZRI S. FRAENKEL[1]

ABSTRACT. A two-person game is defined by placing m stones on distinct vertices $(u_1, \cdots, u_m) = u^m$ of a finite directed loopless graph, which may contain cycles. A move consists of moving a stone from a vertex u_i to a neighboring vertex u_j along a directed edge. If there was already a stone at u_j, both stones get annihilated. The player making the last move wins. If there is no last move, the game is a tie.

Let G denote the generalized Sprague-Grundy function. For an arbitrary game position u^m, suppose that $G(u_i) = \infty$ $(1 \leqslant i \leqslant n)$, $G(u_i) < \infty$ $(n < i \leqslant m)$. It is proved that

$$G(u^m) = G(u^n) \oplus \sum_{i=n+1}^{m}{}' G(u_i),$$

where $\oplus$ and Σ' denote generalized Nim-addition. This completely determines the game's strategy.

1. **Introduction and preliminaries.** Throughout, R denotes a finite directed loopless graph, which may contain cycles. A two-person game $C(M)$ is defined by placing m stones on m distinct vertices of R. A move of $C(M)$ consists of moving a stone from a vertex u to a neighboring vertex v along a directed edge. For $m > 1$, there is an additional annihilation rule: If there is one stone at u, the move of an additional stone to u results in the annihilation of both stones. The players play alternately, and the player making the last move is the winner. If there is no last move, the game is a tie.

The purpose of this work is to present a strategy for the game and its disjunctive compounds. The problem was presented to me by Dr. John H. Conway. Detailed proofs will be presented elsewhere.

For any impartial game M of bounded play, a finite *game-graph* R is defined, whose vertices are the game's positions, and (u, v) is an edge directed from u to v if and only if there is a move from position u to position v. As usual, we shall

---

AMS (MOS) subject classifications (1970). Primary 05−00, 05C20, 05C99, 90−00, 90D99.

[1]Supported in part by the National Science Foundation.

occasionally identify M with R and the positions and moves of a game with the vertices and edges of its corresponding game-graph, using them interchangeably.

In order to present the main result in a precise form, a few definitions are required.

I. THE GENERALIZED SPRAGUE-GRUNDY FUNCTION. By $V(R)$, $E(R)$ we denote the vertex and edge sets of R respectively. If $u, v \in V(R)$ are neighboring vertices, we denote by (u, v) the edge directed from u to v. For $u \in V(R)$, the set of *followers* of u is defined by

$$F(u) = \{v \in V(R) | (u, v) \in E(R)\}.$$

For any real-valued function g, we further define

$$g(F(u)) = \{g(v) < \infty | v \in F(u)\}.$$

Let K be a finite set of nonnegative integers, $M(K)$ the smallest nonnegative integer not in K. The Generalized Sprague-Grundy (GSG) function maps the vertices of R into the nonnegative integers and ∞, subject to **A, B, C** below. If $G(u) = \infty$, we often use the notation $G(u) = \infty(L)$, where $L = G(F(u))$. If all followers of u are labeled ∞, then $L = \emptyset$. Further, instead of $M(G(F(u)))$ we shall use the simpler notation $M_G(u)$.

A. If $G(u) < \infty$, then $G(u) = M_G(u)$.

B. $G(u) = \infty$ if and only if there exists $v \in F(u)$ satisfying $G(v) = \infty(L)$, $M_G(u) \notin L$.

C. For every set of vertices admitting both assignments **A** and **B**, assignment **B** is made.

In [1] we showed that G exists uniquely for every R. See also [2].

II. DISJUNCTIVE COMPOUNDS. As is customary, we denote by N the set of all positions for each of which there is a move by which the *next* player can force a win, no matter what his opponent may do. By P we denote the set of all positions such that if the previous player leaves his opponent in one of them, the *previous* player can force a win, no matter what his opponent may do. Finally, we denote by T the set of all vertices which are tie positions, i.e., positions from which no player can force a win.

A *disjunctive compound* of $m \geq 1$ games is a two-person game $D(M_1, \cdots, M_m)$ composed of a set of m games $M_1, \cdots, M_m$. Each player at his turn chooses one game and makes a move in it. The first player able to move in no game is the loser, the other the winner. If there is no last move, the game is a tie.

If the m games are identical, i.e., $M_i = M$ $(1 \leq i \leq m)$, the disjunctive compound $D(m)$ can be described by placing m stones on any (not necessarily distinct) vertices of the game-graph R of M. Each player at his turn selects one stone and moves it to a neighboring vertex along a directed edge—but without annihilation. Described in this manner, a certain similarity between the disjunctive compound of a game and an annihilation game becomes apparent.

The game-graph $D(R_1, \cdots, R_m)$ of the disjunctive compound of m games is defined by $R_1 \times R_2 \times \cdots \times R_m$, where R_i is the game-graph of M_i $(1 \leqslant i \leqslant m)$. Its vertices are given by

$$u^m = \{(u_1, \cdots, u_m) | u_i \in V(R_i)\}, \qquad v^m = \{(v_1, \cdots, v_m) | v_i \in V(R_i)\},$$

and (u^m, v^m) is an edge if and only if there exists j such that $(u_j, v_j) \in E(R_j)$, and $u_i = v_i$ for all $i \neq j$. Define $\sigma(u^m) = \Sigma'_{i=1}^{m} G(u_i)$, where Σ' is the ordinary Nim-sum (binary addition without carries) if $G(u_i) < \infty$ $(1 \leqslant i \leqslant m)$, and the additional provisions

$$a \oplus \infty(L) = \infty(L \oplus a) \quad (L \oplus a = \{l \oplus a | l \in L\}),$$

$$\infty(K_1) \oplus \infty(K_2) = \infty(\varnothing).$$

Now,

THEOREM I. $G(u^m) = \sigma(u^m)$ *for all* u^m.

THEOREM II.

$$P = \{u^m | G(u^m) = 0\},$$

$$N = \{u^m | 0 < G(u^m) < \infty\} \cup \{u^m | G(u^m) = \infty(L), 0 \in L\},$$

$$T = \{u^m | G(u^m) = \infty(L), 0 \notin L\}.$$

(See [1], [2].) These two theorems show that the GSG-function completely determines the strategy of the disjunctive compound, and, in particular, of a single game $(m = 1)$.

2. The main result.

The game-graph $C(R)$ of $C(M)$, to be called the *contrajunctive compound* of M, has vertices of the form $u^m = (u_1, \cdots, u_m)$, $u_i \in R$, $u_i \neq u_j$ $(1 \leqslant i < j \leqslant m)$, and edges (u^m, v^n), where $v^n = (v_1, \cdots, v_n)$, with $(u_j, v_j) \in E(R)$ for some j, and $u_i \neq v_i$ for all $i \neq j$. We use the notation $C(R) = R \otimes \cdots \otimes R$ (m times). Note that v^n can have two forms:

(i) $v_j \neq u_i$ $(1 \leqslant i \leqslant m)$. Then the move involves no annihilation, $n = m$, and

$$v^n = (u_1, \cdots, u_{j-1}, v_j, u_{j+1}, \cdots, u_m).$$

(ii) $v_j = u_i$ for some i. Then annihilation takes place, $n = m - 2$, and

$$v^n = (u_1, \cdots, u_{i-1}, u_{i+1}, \cdots, u_{j-1}, u_{j+1}, \cdots, u_m).$$

DEFINITION. We say that u^m is *of type* n $(0 \leqslant n \leqslant m)$ if $G(u_i) = \infty$ $(1 \leqslant i \leqslant n)$, $G(u_i) < \infty$ $(i > n)$.

Our main result follows.

THEOREM III. *Suppose that* $u^m \in V(C(R))$ *is of type* n. *Then*

(1)
$$G(u^m) = G(u^n) \oplus {\sum_{i>n}}' G(u_i).$$

3. Ramifications and indications of proof. A move (u_j, v_j) in the contrajunctive compound can be classified into one of three categories:

A *type-preserving move*, where either $G(u_j) = G(v_j) = \infty$ and no annihilation takes place, or else $G(u_j) < \infty$, $G(v_j) < \infty$.

A *move changing type by* 1, where either $G(u_j) = \infty$, $G(v_j) < \infty$ or vice versa. (Annihilation may take place.)

A *move reducing type by* 2, where $G(u_j) = G(v_j) = \infty$, and annihilation takes place.

Note that every move in the disjunctive compound precipitates a change in precisely one of the summands constituting the Nim-sum. For a contrajunctive compound, on the other hand, a move changing the type by 1 induces a change in *both* summands on the right of (1). The remarkable point to note is that the Nim-sum determines the strategy even in this case.

Suppose now that the initial position u^m of $C(M)$ is of type n. Then the analysis of $R^n = R \otimes \cdots \otimes R$ (n times) completely determines the game's strategy in the following sense: Suppose that $u^m \in N$. The proof of Theorem III shows that there exists $v^h \in F(u^m) \cap P$, v^h of type $\leqslant n$. Also, if $u^m \in P$, then even if $v^h \in F(u^m) \cap N$ is of type $n + 1$, there exists $w^k \in F(v^h) \cap P$ such that w^k is of type $\leqslant n$. Finally, if $u^m \in T$, then even if $v^h \in F(u^m) \cap T$ has type $n + 1$, there exists $w^k \in F(v^h) \cap T$ such that w^k is of type $\leqslant n$.

Suppose that R contains k vertices labeled ∞. The analysis of R^n ($n \leqslant k$) involves $K = \sum_{i=0}^{n} \binom{k}{i}$ vertices, and the computation of G involves $O(K^3)$ operations [1]. Actually, G has to be computed and stored only for all vertices u^j of type j $(0 \leqslant j \leqslant n)$. The analysis of R^n permits playing a game with any initial position of type $\leqslant n$. If it is desired to analyze every game with arbitrary initial position, then the graph R^k has to be analyzed. The computation of G for R^n can be done conveniently by computer. Theoretical results about $G(u^n)$ as a function of R^n will be included together with the detailed proofs.

Let ρ be the maximal out-degree of a graph R with k vertices. Using ρ counters $C_1, \cdots, C_\rho$, which are all initially cleared to zero, we present an algorithm [3] for computing G (see also [1], [2]):

(1) If there exists an unlabeled $u \in V(R)$ with $M_G(u) = i$ such that every unlabeled $v \in F(u)$ has a follower w labeled i, then put $C_i + 1 \rightarrow C_i$ and label u with (i, C_i). Return to 1.

(2) Label all unlabeled vertices with ∞.

The validity proof of this algorithm—which also requires $O(k^3)$ steps—is an immediate consequence of the definitions **A, B, C** above, and the uniqueness of G.

The strategy for any disjunctive or contrajunctive compound is controlled by the counters C_i: If a certain move from N to P requires going from u_i to v_i with $G(v_i) = j$, then that follower v_i is chosen whose associated C_j has minimal value. This guarantees winning in a finite number of steps.

The proof of Theorem III is by induction on the type n, with induction basis $n = 0$. Denote by G' the right-hand side of (1). If $G' \neq G$, let t be the minimum value of G' for which $G' \neq G$. Then it can be shown that there exists a cycle C on which G assumes a minimum value t, with the additional property that $u \in C$, $G(u) > t \Rightarrow$ all $v \in F(u)$ with $G(v) = t$ are also on C. But if w is the first vertex of C labeled by the above algorithm, it clearly cannot be labeled by any finite value. This contradiction shows that $G' = G$.

REFERENCES

1. A. S. Fraenkel and J. Perl, *Constructions in combinatorial games with cycles*, Proc. Internat. Colloq. on Infinite and Finite Sets, Keszthely, June 1973 (to appear).

2. C. A. B. Smith, *Graphs and composite games*, J. Combinatorial Theory **1** (1966), 51–81. MR 33 #2572.

3. ———, *Simple game theory and its applications*, Bull. Inst. Math. Appl. **7** (1971), 3–7.

WEIZMANN INSTITUTE OF SCIENCE

Proceedings of Symposia in Applied Mathematics
Volume 20
1974

THE INTEGRATION OF COMPUTING AND MATHEMATICS
AT THE OPEN UNIVERSITY (AN ABSTRACT)

BY

F. B. LOVIS AND R. V. M. ZAHAR

Introduction. The Open University is a British university, founded by Royal Charter in May 1969, which began its teaching in January 1971.

Whilst the University is organizationally similar to the other United Kingdom universities, operationally it is very different. The employment of broadcasting, technological devices and correspondence tuition enables it to teach the student in his own home and therefore, unlike any other British university, it has no resident student population.

In a typical week's work the student of a whole credit course (32 units) is expected to work for some ten hours. He will have a substantial piece of correspondence material to study, together with its associated exercises, will watch a television programme of 25 minutes and listen to a radio programme of 20 minutes duration. He will attend personal tutorials on a more or less regular basis, according to his personal wishes and will be required regularly to submit tutor marked and computer marked assignments. A student of a half credit course (16 units) will follow the same pattern but will have a fortnight in which to do each unit of ten hours work.

This system of teaching enables tuition to be offered to all persons living in the United Kingdom and amongst the registered student population of almost 40,000 there are those who live in places as far apart as Cornwall and the Shetland Isles. Most of the students are in full time employment and read for a degree in their spare time. From the outset, the University has attracted many who, through a variety of circumstances, social deprivation, economic hardship or military service, were deprived of the opportunity to pursue a traditional University education.

AMS (MOS) subject classifications (1970). Primary 68A10, 98—00, 98A10, 98A30, 98A35.

Degree courses are offered at both ordinary and honours levels and suitably qualified persons may read for a higher degree. Broadcast media is produced by the BBC for the Open University at Alexandra Palace and is transmitted on radio and television national networks.

The teaching system. Teaching is mainly by correspondence tuition which is supplemented by broadcast lectures and demonstrations, backed by face to face tuition, student counselling and annual residential summer schools. It is the combination of all these elements, together with the local support services, which makes the Open University unique.

In servicing a student population of almost 40,000 throughout the United Kingdom it is inevitable that a number are not able to receive broadcasts either by radio or television; others may be the victims of temporary technical faults or unable to meet published transmission times, whilst their course moves inexorably forward. This, together with the stated aim of the system to provide face to face tuition and counselling (the main local support services), has led the University to establish 280 Regional Study Centres throughout the United Kingdom. It is at these centres that most support services are made available to the student. A comprehensive library of course media, television films, radio tapes and course units is maintained at each centre, together with television and radio replay facilities. The Study Centres are attended by the academic staff of the University and it is at these centres that the student receives face to face tuition and counselling from his appointed tutor. They also provide a useful forum where students can meet each other, engage in discussions, consider mutual problems and generally avail themselves of the atmosphere and stimulus so often taken for granted at traditional universities.

The undergraduate courses are structured in such a way that a student requires only six full credits for a Bachelor's degree and eight for an honours. At present, the Mathematics Faculty offers three full-credit courses and six half-credit courses, with plans to extend the total number of half-credit courses to approximately 15. Each of the existing courses has a component which involves computing aspects and it is expected that the majority of the remaining courses will also have such a component. Specific computing exercises are set in most mathematics courses and, in addition, students are encouraged to use the computer to provide an intuitive idea of how to solve problems in general.

The Student Computing Service. The Student Computing Service, which was established to provide computing facilities for all students of the University, owns 3 Hewlett-Packard 2000B computers which are situated in London, Newcastle (in the northeast of England) and on the University's campus at Milton Keynes (in central England).

It is significant that 30% of the backing store is given over to system use and 30%–40% to the system program library. The remaining 30%–40% is not sufficient to allow the students to save their programs and it is accordingly hoped that this restriction will be removed by a change to the HP2000F system, which allows virtually unlimited backing store.

The only language available to students is H.P. BASIC but this is not regarded at present as a serious restriction since it is thought that this is the most suitable current language for our method of operation.

The student uses the Student Computing Service in one of two ways. The great majority make interactive use of one of 192 terminals which are situated in some 160 locations throughout the United Kingdom and Northern Ireland. Nearly all these locations are in the University's study centres and the terminals are available for use from 6:00–10:00 p.m. on each weekday, and on Saturday mornings.

The students use the terminals without supervision and without any resource to advice, other than what may coincidentally be available. The Student Computing Service must therefore depend upon the students themselves to report any malfunction of the equipment. In instances where this is not done, there is no way of preventing disappointment for the students who have booked the terminal on the following evening.

There is no doubt that the use of the service involves the students in some trouble. Few of them live close enough to a terminal to go along on the off-chance of finding it free. The great majority need to book their sessions in advance and are easily dissuaded by any news of malfunction or down-time. It is accordingly necessary to convince the student that he will receive a definite advantage, especially in his attempt to gain credit for the course, from his work at a terminal. This, in turn, seems to depend on the extent to which the computing element has been integrated within the particular course. In M201, Linear Mathematics, the computing element is not an integral part of the course and only 10% of the students bother to do it. A slightly higher degree of integration has been achieved in the Foundation Course, M100, and 50% of its students use the system. In two computing courses, M251, An Algorithmic Approach to Computing, and PM951, Computing and Computers, 100% of the students are naturally involved. Precise information about the amount of use of the Student Computing Service by each student is not yet available.

Since it cannot be assumed that every student will be able to get to a terminal, the Student Computing Service offers a back-up postal service. This offers a one-day turn around but is, in general, used by surprisingly few students. The exceptional course in this respect is M251 which supplies 300 of the 500-odd students at present using this facility. The method used is for an operator at the installation at Milton Keynes to examine the coding sheets, to correct any obvious errors, and then to run the program at a terminal. The student is then sent the listing, the print-out and the operator's comments.

During the day-time the Student Computing Service is available for all members of the University staff and computer time is also sold to a large number of external users, both educational and commercial.

Use of terminals. The amount of use varies tremendously from terminal to terminal. At the end of each year the Student Computing Service considers carefully which terminals should be re-sited but there seems to be no way of providing a national network without placing some terminals in places where they receive little use.

The use made of the system by the students of the various courses also varies tremendously from month to month. In June 1973, 80% of the students of the Technology Foundation Course used the service, but only 3% of the students of MST281, Elementary Mathematics for Science and Technology, did so. Our ideal of steady utilization of the resources throughout our academic year seems unlikely ever to be approached.

Computer usage.

Mathematics Foundation Course. In 1972 a random sample of 1600 students was invited to complete and submit report forms on each unit of work. 74% of these did return one or more forms and examination of their replies shows some interesting indications.

The computing element in M100 occurs in Units 8 and 20 (specifically devoted to computing); 26 and 28 (Linear Algebra 3 and 4); 34 (Number Systems) and in the Additional Computing Exercises.

Units 8 and 20 were each intended to occupy the students in 10 hours study, not including the computing activities which were specified. The Additional Computing Exercises were an entirely optional activity, provided because of the long gap between the two computing units and intended merely to prevent the student from getting too much out of practice.

In Unit 26 (Linear Algebra 3) a library program was provided to perform row operations to simplify or invert a matrix. Very detailed instructions were given for the running of the program.

In Unit 28 (Linear Algebra 4) a library program was similarly provided, to solve simultaneous equations by iteration. On this occasion, the students were not merely asked to use this program but first to improve it (i.e. to ensure convergence) and then to run it.

In Unit 34 (Number Systems) they were asked to draw a flow chart for Euclid's algorithm for finding the greatest common divisor of two numbers and then to write and use a corresponding program. Alternatively, they could simply copy the solution which was given.

As will be seen from the figures below, there was a great difference in the number of students who attempted these pieces of work. It seems reasonable to suppose that the students were much more attracted to the type of activity in Unit 26, where they were given a lot of help in using the computer as a tool.

In any one "average" week, about 60% of the students who sent in the returns would have done some computing. Thus: Unit 8 was worked on by 34%; 20 by 19%; 26 by 12%; 28 by 1%; 34 by 2% and the Additional Computing Exercises by 8%. These figures are calculated from the returns for the whole year and ignore the sequential nature of the course.

Some further simple statistics are given below. Bearing in mind that the students of this course had expressed no wish to do computing, and that the course team was aiming to produce two units of equal length and difficulty, it is interesting to observe that the students tackled this topic with a lot of determination and interest. It is clear that in general the students found Unit 8 both unexpectedly easy and short, while Unit 20 is harder and longer than it ought to be. But although it takes so long to do Unit 20, the proportion who find it very interesting is remarkably high. According to the returns, Unit 20 is in fact the longest— and this means *excessively* long—of *all* the 32 units of the course, yet only five units receive a higher rating for "very interesting."

Degree of Difficulty	All Units	Unit 8	Unit 20
Very or fairly difficult	56%	34%	63%
Just right	28%	35%	25%
Degree of Interest			
Very interesting	33%	43%	42%
Fairly interesting	51%	47%	43%
Time Spent on Unit (Target time: 10 hours)			
Up to 8 hours	22%	36%	16%
Over 18 hours	13%	14%	29%

M251—*An Algorithmic Approach to Computing*—(a second-level course) could almost equally well be called a Computing Approach to Algorithms. It is an introductory course to computing which is available to students irrespective of whether or not they have taken the Foundation Course in Mathematics and which demands very little mathematical knowledge.

The theme of the course is problem solving in general, but it concentrates on those problems which lend themselves to a systematic or algorithmic solution which can be obtained by the use of a computer. It is *not* a course on programming.

The course falls into several more or less distinct sections. Of its 16 units, the first 6 concentrate on the stages involved in solving a problem by the use of a computer. The student has a lot of practice in formulating algorithms, expressing them in flow chart notation and writing suitable computer programs. By the end of Unit 6 quite difficult algorithms are being dealt with.

The early exercises are mainly arithmetical, and include: evaluation of formulae; counting; print-out of scores of simple games; giving change for a bill; summation of squares, cubes, etc.; calculation of parcel postal charges; generation of magic squares; calculation of mean and median; Sieve of Eratosthenes; Knights move; calculation in 3 digit-floating-point arithmetic.

By Unit 6 the student moves on to such exercises as: reversing order of a string; construction of subroutines; encoding and decoding messages; searching for a pattern in a string.

Unit 7, on Computer Hardware, stands by itself and displays the computer as a logical machine, showing how it executes instructions. It is followed by 4 units about the algorithms which are used to handle data on a computer. These units show how computers are used to tackle nonmathematical problems and introduce the students to some of the ideas of commercial data processing. In these units, the problems include: access vector representation of connectivity matrix; insertion and deletion in linked lists; construction of logical trees; writing a program to solve Tower of Hanoi; insertion of new record into a queue; file update program; data vet program; magnetic tape merge sorting algorithms; sorting list with use of access vector; comparison of efficiency of hash coding rules; setting up indexed file of variable length records; deletion of record with given key from indexed sequential file.

Units 12 to 16 look at software, paying special attention to programming languages and compiling. Exercises include: interchanging two strings held as lists; deleting all characters contained in one given string from another given string; conversion from infix to postfix form; compiling a specially designed high-level programming language, ACE; tracing operation of multiprogramming supervisor; construction of service routines.

It will be seen that M251 contains little significant mathematical content and a great many computing techniques. It starts, as do many such introductory courses, with arithmetical algorithms because it is easy to set simple computing exercises involving the basic arithmetic operations. When the exercises become harder, they cease in general to be mathematical.

Elementary numerical analysis in the Foundation Course. Because each full credit course at the Open University is meant to be equivalent to approximately 15 semester hours, each of our courses must cover a number of areas which, at the more traditional universities, would be presented as separate topics. Consequently, we attempt to integrate the various topics in each course as much as possible, and thus provide a broader view of mathematics. One such topic, elementary numerical analysis, is involved in a significant portion of our Mathematics Foundation course, not only as a legitimate area of mathematics which is deserving of study in its own right, but also as a fund of examples for demonstrating the abstract concepts that the students see in alternate weeks.

In the Foundation Course, the students meet the concepts of interpolation and iteration during the first few weeks and are introduced to BASIC programming and our on-line facilities early in the course. They are encouraged to conduct their own numerical experiments as the course proceeds, in the hope that they will achieve an appreciation of numerical algorithms and an awareness of practical programming. When discussing mathematical constructions the course then draws on the intuition the students have gained from their experiences.

As a particular, and perhaps not so obvious, example consider the relative error $r(a \times b)$ in the product of two nonzero real numbers a and b:

$$r(a \times b) = r(a) \,\square\, r(b) \quad \text{where} \quad r(a) \,\square\, r(b) = r(a) + r(b) + r(a)r(b).$$

This is a practical situation which can be used to demonstrate the concept of a morphism r between the groups $(R - \{0\}, \times)$ and $(R - \{-1\}, \square)$ indicated by the following commutative diagram:

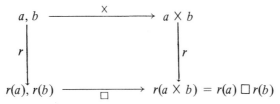

And because the students recognize the practical advantage in predicting $r(a \times b)$, they immediately appreciate the distinction between the two paths from a, b to $r(a \times b)$ in the diagram and the practical reason for following the counterclockwise one.

In a similar manner, one-point iterative methods for the solution of nonlinear equations are used to introduce the ideas of sequence and limit, as well as convergence and divergence. Polynomial interpolation is employed in the development of Taylor series, which in turn are used to demonstrate infinite series and the corresponding calculus.

The virtue of this approach is that numerical analysis and more "pure" mathematics are shown to be perfectly consistent with each other. The students learn the need for a careful mathematical analysis of a problem in conjunction with the application of a numerical algorithm. Moreover, they experience in a natural setting, concepts such as rates of convergence, which are often omitted from early analysis.

Numerical and Linear Mathematics. Our second level course in Linear Mathematics has a large numerical component, which includes the analysis of recurrence relations, numerical solutions of linear equations, numerical solutions of ordinary differential equations, Chebyshev expansions and the numerical eigenvalue problem. These subjects are easily combined with the more theoretical aspects of vector spaces, matrices and linear transformations, the resulting advantage being that the students are given a treatment of all phases in the solution of the corresponding problems.

Although the mathematics of recurrence relations can be quickly described in terms of linear transformations on the vector space of complex sequences, and although their solutions by straightforward recurrence from initial values is trivial in principle, they can suffer from catastrophic numerical instability. In this course, well-chosen examples lead naturally to the study of inherent and induced instability, and to the introduction of condition numbers. Backwards recurrence and the resulting analysis is described and, for perhaps the first time in their careers, the students come to appreciate the interaction between practical computing and mathematical theory.

For systems of linear equations, Gaussian elimination and its mathematically equivalent forms are developed via their theoretical connection with the Hermite normal form. For the discussion of errors in the numerical solution of equations, backwards error analysis is introduced, and then pivoting and iterative refinement are discussed. In a similar manner, various eigenvection techniques are presented at much the same time as the corresponding theory is developed. Our comprehensive approach to the solutions of such linear problems avoids the undesirable impression often gained by students, namely, that matrix theory is fragmented and specialized.

A unified approach to partial differential equations. Although the combination of numerical and linear algebra in a single course may be regarded as a pleasant luxury, the corresponding combination in a course on partial differential equations is almost a necessity. So very few of the partial differential equations which arise in practice possess an analytical solution, that this points out the need for numerical considerations. At the same time, however, it is important in attempting to solve such an equation numerically, that one is cognizant of any relevant theory and that one uses this theory to best advantage in devising the numerical model.

Despite the lack of textbooks which provide an effectively commixed approach to partial differential equations, our third level course on the subject attempts to achieve this. The nature of the course is demonstrated by two of the case studies presented in it: one concerned with manometer response time, the other with overhead wires on high speed electric trains.

In the manometer problem (Jones and Jordan [3]) it is desired to find the time taken for small air pressure differences to transmit over long distances through small bore tubing. As a function of the distance along the tubing and of time, the pressure p satisfies the heat conduction equation in its nonlinear form:

$$\frac{\partial p}{\partial t} = \frac{a^2}{8\eta} \frac{\partial}{x} \left(p \frac{\partial p}{\partial x} \right).$$

In practice, however, this equation is usually linearized, and a theoretical solution found. The effect of such linearization can be shown to be negligible by comparing the results achieved from the linear theory with those obtained in laboratory experiments. But in the context of our course, this negligibility for given values of the parameters is asserted by comparing the theory with the finite-difference solution of the full nonlinear problem found by the Crank-Nicolson method.

In a second case study, the motion of the overhead wire is analysed, this motion resulting from the upward force F of the moving train's pantograph (Gilbert and Davies [1]). This is described by a wave equation containing several force terms:

$$- T \frac{\partial^2 y}{\partial x^2} + \rho \frac{\partial y^2}{\partial t^2} = F(x, t) - \eta \frac{\partial y}{\partial t} - \rho g - y S(x).$$

The solution of this problem is developed by considering several partial problems in order of their increasing complication. First, the static wire in a medium of constant elasticity S is considered and, for reasonable numerical values of the parameters, the effect of gravity g is shown to be negligible. Next, by the method of Fourier series, solutions are obtained for the free oscillation of the wire. The pantograph is then introduced, and another solution is obtained after a change of variable to the frame of reference moving with the pantograph. Finally, a periodic elasticity is introduced, and a solution achieved by a method of perturbation expansions. Comparison with practical experiments shows that the perturbation expansion is invalid, and this points out the need for numerical considerations.

The pantograph example emphasizes the advantages of considering the numerical solution as simply another phase in the analysis of the problem. It would have been unwise for the student to tackle the numerical solution of the original problem without first carrying out the theoretical phases. At the very least, the theory helps the student to gain a physical intuition of the problem and points out the technical sensitivities in the solution. More importantly, the analytical solutions suggest how the numerical problem can be formulated, provide estimates for the length of wire that one should consider, and give an approximate first solution on which one can numerically iterate. In other words, the analytical theory and numerical experiments must be inextricably linked before a realistic solution can be found.

Pure mathematics and computation. Our second-level course in pure mathematics presents the theory of computation as well as set theory, abstract algebra and topology. It was decided to include computation to help the students become aware not only of the mathematical techniques used in the theory of automata, but also of the distinction between constructive and nonconstructive proofs. Given that the students who take this course have also had a prior introduction to practical computing, it was expected that the study of computation would be more relevant to them than another alternative, foundational logic.

The material concerned with computation includes finite-state machines, Turing machines, recursive functions and Post canonical systems. The concepts in the computation part of the course and those in the other parts intersect mainly in set constructions, recursive definitions, induction and function specifications, and this provides a certain degree of continuity in the material. The blending of machine theory with set theory, group, ring and Galois theory and topology has produced some interesting concomitant effects on the attitudes of the students.

Initially, we have found that most students experiencing abstract proofs for the first time maintain a scepticism in their attitude. When they meet existence proofs early in the course, they express a certain degree of dissatisfaction, unless the corresponding construction is also presented. For instance, consider the set-theoretical proof for the existence of the greatest common divisor of two integers a and b not both zero, which is based on noting that the set

$$M = \{na + mb: n, m \in Z\}$$

contains a smallest positive integer. Most students would say initially that the proof is incomplete unless the corresponding Euclidean algorithm for finding the g.c.d. is also included.

Nevertheless, as the course progresses, the students become accustomed to the more nonconstructive proofs and gradually achieve the facility of creating such proofs on their own. At the same time, however, they learn to appreciate the algorithmic content of the proofs that occur in the study of machine theory. It is gratifying to note that towards the end of the course, they express some surprise at discovering that the set of partial recursive functions from the set of natural numbers ω, to itself, is countable, whereas, by Cantor's theorem, the total number of functions from ω to ω has power $2^{\aleph_0}$. As a result of the integration of disciplines in this course, they have become sensitive, at a very early stage in their careers, to relatively profound philosophical questions.

At the very least, the course encourages an appreciation of practical algorithms and their construction in contexts where they are not normally presented. As a result it is not uncommon to find students thinking beyond the usual abstract treatment of factorizing polynomials over the rationals, and asking for practical methods of achieving such factorizations. At such a juncture, they are referred to an area of current activity in computer research (see for example, Knuth [2]).

We have not attempted here to give a comprehensive treatment of our mathematics courses. Rather, through our illustrative examples, we have endeavoured to present the flavour of our approach to the integral nature of computing and mathematics.

REFERENCES

1. G. Gilbert and H. E. H. Davies, *Pantograph motion on a nearly uniform railway overhead line*, Proc. IEEE 113 (1966), 485–492.

2. D. E. Knuth, *The art of computer programming*, Vol. 2: *Seminumerical algorithms*, Addison-Wesley, Reading, Mass., 1969. MR 44 #3531.

3. C. Jones and D. W. Jordan, *Time lags in the transmission of pressure disturbances along long lengths of small bore tubing*, British J. Appl. Phys. 13 (1962), 420–423.

THE OPEN UNIVERSITY

Proceedings of Symposia in Applied Mathematics
Volume 20
1974

REAL TIME COMPUTER GRAPHICS
TECHNIQUES IN GEOMETRY

BY

THOMAS BANCHOFF AND CHARLES STRAUSS

This paper contains three geometric examples, each of which could be given as an exercise to an undergraduate. Like most exercises, they represent illustrations of some mathematical phenomenon which had been observed by a teacher or textbook writer. What unifies the three and makes them noteworthy is that each was first observed by the authors while interacting at a graphics terminal with pictures produced by a computer. Once the observation has been made, it has not been difficult to write out the verification, so far anyway. The hope is that soon such methods will lead to more subtle observations and ultimately to some significant mathematical results. For now, these examples may serve as a demonstration of the power of computer graphics in suggesting geometric problems and in aiding in their solutions.

Example. The evolute of the cardioid. A graphics routine for parallel curves to curves in the plane takes in a curve in parametric form $X(t), a \leqslant t \leqslant b$, and a number n of equal subdivisions of the parameter domain and displays the parallel curve

$$Y_r(t) = X(t) + rN(t)$$

where the oriented distance r is put in by a control dial. The unit normal $N(t)$ is either given by a formula or is obtained by taking the average of the normal vectors to the edges adjacent to the point $X(t)$ in the polygonal approximation determined by the partition.

If $X(t)$ is smooth, and if $X'(t) \neq 0$ for all t, then, for sufficiently small r, the parallel curve $Y_r(t)$ is also smooth; but, for large r, the curve may develop cusps:

AMS (MOS) subject classifications (1970). Primary 53–04, 68A40.

$$Y_r'(t) = X'(t) + rN'(t)$$
$$= X'(t) + r(- \kappa(t)X'(t))$$
$$= (1 - r\kappa(t))X'(t).$$

We get a cusp only when $Y_r'(t) = 0$, i.e., when $r = 1/\kappa(t)$. The locus of cusps of parallel curves is called the *evolute*, so the equation of $E(t)$, in the smooth case for $\kappa(t) \neq 0$, is given simply by

$$E(t) = X(t) + (1/\kappa(t))N(t).$$

We may also describe the evolute as the envelope of the normal lines, so that $E(t)$ is the limit as $h \longrightarrow 0$ of the intersection of the line $X(t) + uN(t)$ and the line $X(t + h) + uN(t + h)$. In a visual display which includes the segments from $X(t)$ to $X(t) + rN(t)$ for selected points along the parameter domain, the evolute appears as the fold curve where this strip intersects itself.

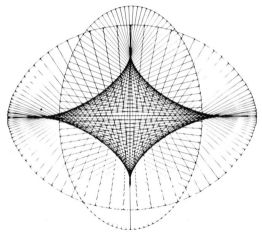

FIGURE 1

The visual display can suggest results which might not be apparent from the equations themselves. After an observation has been made, it is often straightforward to make the verification.

The cardioid is given in parametric form by

$$X(t) = (1 + \cos(t))(\cos(t), \sin(t)), \qquad 0 \leqslant t \leqslant \pi,$$

with tangent vector $X'(t) = (- \sin(t) - \sin(2t), \cos(t) + \cos(2t))$ which is nonzero except at the point $t = \pi$ where the curve has a cusp. The computer picture of the curve plus its normal lines suggested that the evolute itself was similar to the original curve, although rotated, dilated and translated. (See Figure 2.) Once this observation was made, it was not difficult to check.

By standard formulas of elementary calculus, we find

$$\|X'(t)\| = (2 + 2\cos(t))^{\frac{1}{2}}, \qquad \kappa(t) = \frac{3}{2}(2 + 2\cos(t))^{-\frac{1}{2}},$$

$$N(t) = (2 + 2\cos(t))^{-\frac{1}{2}}(-\cos(t) - \cos(2t), -\sin(t) - \sin(2t)).$$

Thus

$$E(t) = X(t) + (1/\kappa(t))N(t)$$

$$= \left((1 + \cos(t))\cos(t) - \frac{2}{3}(\cos(t) + \cos(2t)),\right.$$

$$\left. (1 + \cos(t))\sin(t) - \frac{2}{3}(\sin(t) + \sin(2t))\right)$$

$$= \left(\frac{1}{3}(\cos(t) - \cos^2(t)) + \frac{2}{3}, \frac{1}{3}(\sin(t) - \sin(t)\cos(t))\right)$$

$$= \frac{1}{3}(1 - \cos(t))(\cos(t), \sin(t)) + \left(\frac{2}{3}, 0\right).$$

Thus the evolute of the cardioid is another cardioid, scaled down by 1/3, rotated, and shifted by (2/3, 0).

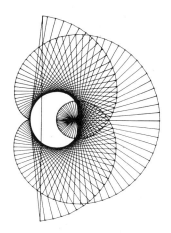

FIGURE 2

For a general discussion of such evolute curves, see for example the book of Lockwood [2].

2. Example. The graph of the complex exponential and its inverse. As part of a program to demonstrate the usefulness of the computer graphics routine in studying surfaces in 4-space, an investigation was carried out of the graph of $w = e^z$ in complex 2-space viewed as real 4-space. Letting $z = x + iy$ and $w = u + iv$, we obtain the locus

$$(x, y, e^x \cos(y), e^x \sin(y)).$$

The domain used was $-2 \leqslant x \leqslant 2$, $-2\pi \leqslant y \leqslant 4\pi$, cross-hatched by choosing equal subdivisions in the x- and y-domain. Projection to the (x, y, u) space gives the graph of the real part of e^z as a surface in 3-space, and projection to (x, y, v) gives the graph of the imaginary part. A smooth rotation in the u-v-plane through angle α corresponds to the graph of

$$(x, y, e^x(\cos(y)\cos(\alpha) + \sin(y)\sin(\alpha)))$$

$$= (x, y, \operatorname{Re}(e^{x+iy})e^{i\alpha}) = (x, y, e^x e^{i(y+\alpha)}).$$

A rotation of this locus into (x, u, v) space produced a surface which should have been expected but which as a matter of fact appeared as a surprise. Under a rotation in the y-v-plane, the surface wraps itself around into a surface of revolution, an exponential horn with equation

$$(x, e^x \cos(y), e^x \sin(y)).$$

(See Figure 3.)

FIGURE 3

A further rotation in the x-y-plane brings the surface into a right conoid

$$(y, e^x \cos(y), e^x \sin(y)).$$

(See Figure 4.)

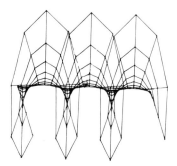

FIGURE 4

We can identify the significance of these last two projections of the surface in 4-space by reparametrizing in terms of u and v. The first becomes

$$(\ln((u^2 + v^2)^{1/2}), u, v)$$

and the second becomes

$$(\tan^{-1}(v/u), u, v).$$

But the functions $\ln((u^2 + v^2)^{1/2})$ and $\tan^{-1}(v/u)$ are precisely the real and imaginary parts of the function $\ln(w)$. This should have been anticipated since the locus (z, e^z) viewed another way gives the locus $(\ln(w), w)$. As in the case of a curve graphed in the Euclidean plane, an accurate graph of a function can be looked at from a different point of view to give the *inverse* function.

The domain of the inverse function $\ln(w)$ here is given by the image of the function e^z, where we have

$$e^{-2} \leqslant |w| \leqslant e^2 \quad \text{and} \quad -2\pi \leqslant \arg(w) \leqslant 4\pi,$$

a multiply covered domain in the w-plane.

3. Example. Deformation of a cusp of an algebraic curve. An algebraic curve in complex 2-space may be described by a relation between variables z and w, for example, $z^3 = w^2$.

Such a complex equation yields two real equations so the locus can be considered a real surface in real Euclidean 4-space. We may parametrize this locus by choosing a radial domain $\zeta = re^{i\theta}$, $0 \leqslant r \leqslant 1$, $0 \leqslant \theta \leqslant 2\pi$, and taking $z = \zeta^2$, $w = \zeta^3$. We then have a mapping of the disc into complex 2-dimensional space

$\langle r^2 e^{i2\theta}, r^3 e^{i3\theta} \rangle$ or, in real coordinates,

$$(r^2 \cos 2\theta, r^2 \sin 2\theta, r^3 \cos 3\theta, r^3 \sin 3\theta).$$

An extensive investigation of the singularities of the projections of this surface has been carried out and reported in [1], and during this investigation one projection developed which had not been expected. Rotation in the u-y-plane by $45°$ followed by rotation in the x-v-plane by $45°$ brings the surface into the position illustrated below, a projection with a five-fold symmetry. Closer examination of the equations reveals why such a projection could have been anticipated. The effect of the rotation is to give a projection of ζ to $\zeta^2 + i\bar{\zeta}^3$, i.e.,

$$re^{i\theta} \longrightarrow r^2 e^{i2\theta} + ir^3 e^{-i3\theta} = f(r, \theta).$$

We may then express the five-fold symmetry by demonstrating that $f(r, \theta + 2\pi/5) = e^{i4\pi/5} f(r, \theta)$, i.e.,

$$r^2 e^{i2(\theta + 2\pi/5)} + ir^3 e^{-i3(\theta + 2\pi/5)} = e^{i4\pi/5}(r^2 e^{i2\theta} + ir^3 e^{-i(3\theta + 2\pi)}).$$

FIGURE 5

Conclusion. In each of these examples we have illustrated the process by which a visual representation suggests a geometric result, which then can be proved analytically. Such results hopefully typify a process which will yield many significant results in the future.

The examples here have been illustrated in this article by photographs of displays on a television console. A more effective accompaniment to the article would be a film showing how the interactive graphics routines allow an observer to manipulate the geometric objects in real time.

REFERENCES

1. T. F. Banchoff and C. M. Strauss, *Real time computer graphics analysis of* $z^3 = w^2$ *in Euclidean 4-space*, Talk presented to the Annual Meeting of the Amer. Math. Soc., Dallas, January 1973. (With film)

2. E. H. Lockwood, *A book of curves*, Cambridge Univ. Press, Cambridge, 1971.

BROWN UNIVERSITY

Proceedings of Symposia in Applied Mathematics
Volume 20
1974

VISUAL GEOMETRY, COMPUTER GRAPHICS
AND THEOREMS OF PERCEIVED TYPE

BY

PHILIP J. DAVIS

The inborn capacity to understand through the eyes has been
put to sleep and must be reawakened.
Rudolph Arnheim, *Art and Visual Perception*

This is the Visual Generation, *New York Magazine,* May 28, 1973

ABSTRACT. The author presents arguments in favor of the following two
positions.

(1) Visual geometry ought to be restored to an honored position in mathematics. Computer graphics comprising animation and color offers the possibility of going far beyond conventional drawings.

(2) The classical notions of what constitutes a mathematical theorem or a
mathematical truth need broadening. These notions should be recast so as to include a variety of phenomena which are systematically generated, perceived by
the senses and interpreted by the brain.

1. **Introduction.** It is incumbent upon each mathematician and each generation of mathematicians to formulate a definition of mathematics. Granted that
this is a hopeless task and also granted that no consensus can ever be reached and
that all formulations are evanescent, the exercise is useful in that it compels the
mathematician to think through where he believes his discipline places him in the
world of experience and thought. It is also useful in that it provides future his-
torians of science with a picture of how the past regarded itself.

A popular contemporary mathematical dictionary (James & James) defines
mathematics as "the logical study of shape, arrangement, and quantity." This
definition, unsophisticated though it may be, coincides with the popular view of
what the subject is all about. A definition which goes back a hundred years to
the writings of C. S. Peirce and which emphasizes the logical aspect of the subject

---

AMS (MOS) subject classifications (1970), Primary 69-00, 02-00.

tells us that mathematics is the science of drawing necessary conclusions. An update of the C. S. Peirce definition might be that mathematics is the workings-out of a universal Turing machine. Other contemporaries might talk of mathematics in terms of logical transformations, grammars, invariants, or in terms of structuralism. At the turn of the century, Bertrand Russell, focusing attention on the varieties of external interpretation that one and the same mathematical structure might carry, wisecracked that "mathematics may be defined as the subject in which we never know what we are talking about, nor whether what we are saying is true."

2. **Theorems.** Let us open the average book on mathematics. What kind of thing do we find in it? Well, first of all, we find definitions, theorems and proofs. These constitute the Trinity of contemporary mathematizing and form the hard core of the book. But there may be other things in the book. There may be discussions of the definitions, theorems and proofs. The discussions may be historical or bibliographical or methodological or aesthetic. There may be judgments or indications of where the core material fits in with other mathematics or with other aspects of the universe. Of course, if the book is on applied mathematics then the percentage of this last type of material might very well (but not necessarily) go up.

A book on mathematics might also contain graphical or visual material. This differs from what is found in normal mathematical sentences written in the normal mathematical font of symbols. These are put in by way of elucidation or clarification but are never (by purists anyway) thought to constitute an adequate mathematical proof of anything. There is a widespread feeling that proper proofs can only be carried out in the format canonized by Euclid. This is the mathematical parallel to the feeling of mediaeval theologians that the spirit is pure while the flesh is corrupt; mathematicians are notorious puritans in their own peapatch.

Despite the theorem-olatry of the past several hundred years of mathematics, there is surprisingly little theorem-ology. What is a theorem? How does it operate? What is it for? James & James says that a theorem is a general conclusion proved or proposed to be proved on the basis of certain given assumptions. A somewhat more sophisticated definition of a theorem, adapted from a current book on mathematical logic, goes along the following inductive lines. The axioms of a formal system F are theorems. If all the hypotheses of a rule of F are theorems then the conclusion of the rule is a theorem. The axioms, i. e., the primitive statements or assumptions, are representable as certain strings of atomic symbols. The theorems are representable as certain other strings of atomic symbols. Proving is the process of passing from an axiom string to a theorem string by a finite sequence of allowable elementary transformations. To verify that the next man's putative theorem is, in fact, the theorem he claims it to be is merely to verify that the sequence of string transformations is in order. The whole thing is in principle perfectly mechanizable.

Now that we know what a theorem is, what can we say about theorems in a general way *apart from comments on specific theorems*? Books on mathematics or metamathematics say very little. One authority I consulted told me categorically that the only assertion one would want to make is that a theorem is either true or false (if it is proposed to be proved) or true (if it has been proved), in which case there is no reason at all for mentioning the fact. This is an extreme point of view.

In the mathematical sense one can, e.g., talk about the range of a theorem (whether or not it applies to anything at all) or the generality of a theorem. There is even a recent mathematical theory of the depth of a theorem.

In the extramathematical sense, one can talk about the utility of a theorem, the beauty of a theorem, the popularity of a theorem, the revolutionary quality of a theorem, etc. (A recent mathematical article contains the following sentence: "Theorem 7.4. Hilbert's Tenth Problem is unsolvable!" The exclamation point here is not mathematical notation. Presumably the author is trying to convey to the reader his own sense of elation or surprise at the result.) One can even talk about the possible evolution of the notion of a theorem and not treat the thing as if it were a fixed concept frozen for all future time. There is obviously much that can be said about the theorems in general, although I have the distinct impression that there is a dearth of such talk in the mathematical literature.

3. The visual image. In the early 19th century the greatest accolade that could have been accorded one mathematician by another was to have called him a "geometer." The irony is that at the very time this honorific was in use, the reasons which called it into being were themselves almost dead. The title was a splendiferous archaism.

What are some of the reasons for the decline of the visual image in mathematics?

(1) The tremendous impact of Descartes' *Discours de la Méthode* (1637) by which geometry was reduced to algebra; also the subsequent turnabout wherein the medium (algebra) became the message (algebraic geometry).

(2) The collapse, in the early 19th century, of the view, derived largely

from limited sense experience, that Euclidean geometry has a priori truth for the universe; that it is *the* model for physical space.

(3) The incompleteness of the logical structure of Euclidean geometry as discovered in the 19th century and as corrected by Hilbert an   others (Euclid debugged).

(4) The limitations of two or three physical dimensions 'ich form the natural backdrop for visual geometry.

(5) The limitations of the visual ground field over which visual geometry is built as opposed to the great generality that is possible abstractly (finite geometries, complex geometries, etc.) when geometry has been algebraicized.

(6) The limitations of the eye in its perception of mathematical "truths" (e.g., the existence of continuous everywhere nondifferentiable functions, optical illusions, suggestive but misleading special cases, etc.).

These perceptions and historical developments have been of overwhelming importance. The visual image went into a tailspin from which it has not yet recovered. The little boy played with matches and got his fingers burned, so civilization abolished all the matches instead of training the boy. It is time to restore the image. The image has much that is new to offer. It can be done through the medium of computer graphics.

4. What computer graphics offers. By an interactive computer graphics installation I shall mean—leaving the jargon of computer hardware aside—a television tube hooked up to a computer. This combination is to be addressable by typewriter, lightpen, joystick, control dials or other analogue devices and the whole is to be backed up by sufficient graphics hardware and software that the programming of visual images of the ordinary mathematical variety can be carried out as easily as, say, computation in some well-known interactive languages such as BASIC or APL. Admittedly, at the time of writing (August 1973), this combination is available at very few university computer centers. The availability of really advanced graphical facilities such as color tubes, sketchpads, opportunities for computer animation are correspondingly much more limited.

What are some of the mathematical potentialities of computer graphics?

(1) Insight into situations of a mathematically conventional but possibly difficult nature.

(2) Computer-generated art.

(3) Creation of mathematical theorems of "perceived type."
I shall discuss these points separately.

5. Generation of conventional theorems via graphics. A computer graphics installation can, of course, be used to illustrate a wide variety of principles of elementary mathematics for purposes of instruction. This can be of enormous importance for didactics. Much effort has been spent in the past decade to illustrate various principles of calculus, probability and statistics, mechanics, higher-dimensional geometry, etc. by means of the scope. There have also been illustrations of more advanced things such as mappings induced by analytic functions of a complex variable, certain geometrical principles occurring in the theory of functions of two complex variables such as Bergman's distinguished boundaries, solutions of partial differential equations animated according to the time parameter, the solution of the many-body problems assuming general force laws, studies of singularities of algebraic curves, iterations of nonlinear transformations, projections of higher-dimensional objects and transformations of these objects, etc.

Graphical displays can suggest theorems or truths which the mathematician might then attempt to prove in a conventional way. Conventional proofs of what has in fact been observed may be extremely difficult to obtain. For example, in celestial mechanics one renowned authority (Carl Ludwig Siegel) writes off the possibility of analytic progress in certain areas of the subject. Does this mean that there can be no knowledge in such an area? Nonsense, as any practical man would tell you. Students fooling around with, e.g., orbits in the many-body problem that have been graphically displayed have found periodic solutions whose existence defies our keenest analytical analysis. A systematic graphical exploration of certain topics might lead to a consistent, extensive, interconnected, interesting and important corpus of material which might not have been available through research that is pursued according to the conventional mathematical methodology. To an experimental scientist this point of view is, of course, old stuff. To a mathematical conservative, this might be magnificent *mais ce n'est pas la guerre.*

In the investigation by computer graphics of conventional mathematical problems one also moves to knowledge or experience which I shall call here, for lack of a better term, "theorems or structures of perceived type."

This type of knowledge might be perceived by the individual as a gut feeling. To quote one team of investigators (Banchoff and Strauss):

> Using control dials, joysticks, and other analog input devices, a mathematician can get immediate portrayal of the geometric effect of continuously varying parameters. He also has finger-tip control over the current values and rate of changes of these parameters, encouraging the development of a visceral feeling for the effect of these parameter variations.

This visceral feeling might be of importance in experiences ranging from the highly practical training of aircraft pilots via simulated cockpits to space intuition that might be achieved by moving around objects computer-wise in a higher-dimensional space. The pilot-in-training is learning a body of theorems of "perceived type." There is obviously a close relation here to kinematics and kinaesthetics.

6. **Computer-generated art and animated films.** The variety of output devices in a computer center offers the possibility of computer-generated art and films. The line printer, the plotter, and the scope have all been used; masters of the craft have produced pieces and effects which are nothing short of amazing. At the very lowest level, the computer-driven output device can be regarded as a new medium with characteristic effects, similar to technological processes such as acrylics of silk-screening. Each process has a certain scope and certain strengths and weaknesses. At the very lowest level, computer art might attempt to imitate certain effects obtained by conventional art media. At a higher level, the unique nature of the medium comes into play and one obtains effects which might be difficult or pointless, if not impossible, to create conventionally. At a still higher level, the relationship between the visual effects and the language used to create the effects comes into great prominence. One might even posit an advanced Descartean stage (I have not seen it yet) in which the language turns about and supersedes the visual effect.

Computer art can be carried out for sensual or craft pleasure, for amusement, for aesthetic values, for shock, for practice in programming, or as an adjunct to mathematical investigations of conventional type. It can be carried out for *l'art pour l'art,* or simply because the output devices are there.

I recall seeing Abraham Lincoln's face produced by computer-driven typewriters in the late 40's, done as a demonstration piece for a laboratory "open house." But serious computer art is only about ten years old. It is too early for an iconography to have developed which might lend value independently of the image *qua* image.

On the purely utilitarian level, computer art moves imperceptibly towards commercial art (I have seen some very beautiful stamps with a computer art figure on them) and towards the design of commercial and industrial shapes and thence into the automatic fabrication of such shapes. As such, computer art becomes a genuine topic of applied mathematics.

One paradigm for the production of computer art goes along the following lines. Starting from some mathematical scheme (spirographic geometry, number theory or, for that matter, any illustrable mathematical theory, or digitalized conventional pictures) and employing certain mathematical transformations with considerable parametric freedom and possible even built-in "randomness," one produces output. This output is then monitored and accepted or rejected on the basis of some internalized criterion. This leads to parameter adjustments, program modifications, etc., and a new generation of outputs.

The resulting piece of computer art may very well be accidental or serendipitous in the sense that the artist-programmer may not be able to foresee in advance precisely what will be created, but at the same time it represents a tight control by the artist-programmer over his work in that the output results from a fixed program and is reproducible, given the parameters and the initializing values in the case of a randomizer.

The field of computer art appears to me to be wide open; at the same time, as with all seedlings, its future is moot. I personally feel that the potentialities are much greater with animated images than with static images. I should have liked to have included some instances of animation with my illustrative material, but obviously cannot.

7. Creation of mathematical theorems of perceived type. I come to the nub of my argument. The Cartesian program—i. e., the algebraicization of geometry and of vast portions of mathematics with geometric content—represents a major revolution in the history of mathematics. Nonetheless, as with all revolutions, a certain loss was incurred when the culture of the *ancien régime* was undermined. The algebraicization of geometry must be regarded as a prosthetic device of great power which maps certain aspects of space into analytical symbols. The blind might be unabled to manipulate space through the instrumentality of these symbols, but since one channel of sense experience is denied to the blind, one feels that a corresponding fraction of the mathematical world must be lost to them. Political democracy does not require that all men savor the universe at identical levels of intensity. [1]

The analytic program, then, is a prosthetic device, acting as a surrogate for the "real thing." The unit circle as perceived by the eye and acted on by the brain is a very different thing from the symbol string $x^2 + y^2 = 1$. The two sensations are interrelated and each can be considered as an "analytical continuation" of the other and each is on an even intellectual basis with the other. The eye "perceives" many things about the circle which may be difficult or impossible to mimic via the analytic symbols. The visual circle is the carrier of an unlimited number of theorems which are instantly perceived. The perceived gestalt of the

---

[1] Attempts to translate theorems in one sense perception to theorems in a second sense perception can lead to analytical mathematics of the highest interest and difficulty. See M. Kac [2].

circle is at once the formulation of these theorems and their proof. (In connection with some work on approximation theory, I once had to demonstrate the visually-obvious theorem that a circle cannot be filled up by a finite number of nonoverlapping circles of smaller radius. I was lucky in that I found a simple analytic proof. What if I had been confronted with something as difficult as the Jordan curve theorem and my analytic standards were high?)

The regular isocahedron sitting on my desk and perceived as a three-dimensional object is a different thing from a list of the coordinates of its vertices. It is a different thing from the abstract group of rotations that move it into itself. It is a gestalt, complete in itself, self-vindicating, rejoicing in its uniqueness, the carrier for an unlimited number of "theorems of perceived type" that are grasped or intuited and *do not even have to be stated.*

Chilton's Drawing of {5, 3, 3}

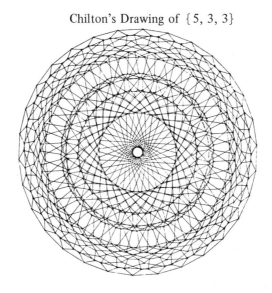

FIGURE 1

(From "Introduction to Geometry", H. S. M. Coxeter, Wiley, 1961)

Take a look at Figure 1. This is a two-dimensional projection of the polytope with Schläfli symbol {5, 3, 3}. The first thing about the figure that catches my eye is that it seems to split up into a number of consecutive rings (at least seven), each of which has a different mesh-pattern or density characteristic. You may seek a conventional proof of this fact if you like, having previously introduced a satisfactory definition of what a mesh-pattern means. I could probably go on for an hour telling what I saw in this fairly complicated image and exceed by far the number of formal theorems in the literature about the polytope $\{5, 3, 3\}$.

Take a look at Figure 2. This was obtained by computing the function $\{|x^3 + y^3| \div 10\} \bmod 3$, and plotting the resultant values 0, 1, 2 as three grey

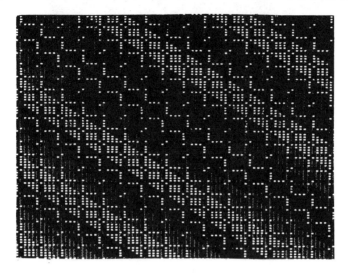

FIGURE 2

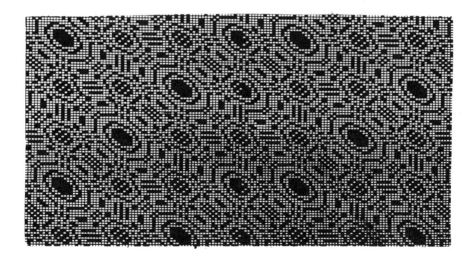

FIGURE 3

Patterns defined by black, grey, and white areas determined by reducing a mathematical function modulo 3. With x, y origin at the center, the top picture is from $\{|x^3 + y^3| \div 10\} \bmod 3$ the bottom one from $[(x^2 + xy + y^2) \div 30] \bmod 3$. (Courtesy K. Knowlton, T. Rainer.)

PHILIP J. DAVIS

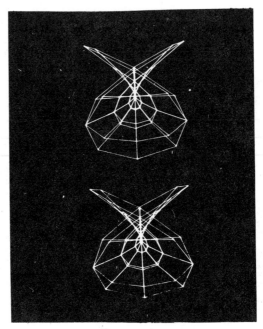

FIGURE 4

Stereo Pair of $(z^2, z^3 + \epsilon z)$

Graphed as a projection of a surface in E^4, with small negative ϵ. (Banchoff & Strauss)

levels; white, grey, black. The resulting herringbone figure of fairly intricate texture, with its symmetries, periodicities, accidentals, is certainly part of the theory
of cubic residues. One might formulate theorems in number theory to account
for what one sees. Some might prove difficult, others trivial. On another level,
though, there is no need for this reduction. One sees what one sees: a characteristic pattern which is the carrier of a mélange of number theorems of the conventional type, but which has an integrity of its own and does not require conventional interpretation. We are seeing a theorem of "the perceived type." In view
of the possibility of such figures, the paucity of geometrical illustrations in books
on number theory is absolutely incredible.

Again my point is not—what we all know—that a good figure can suggest
conventional theorems. It goes beyond. A figure, together with its rule of generation, is automatically and without further ado a definition, theorem and proof
of "the perceived type."

8. What is mathematics? I return at last to the question in the introductory
paragraph. I would suggest that mathematics is the program, the execution, the

output; the gestalt perceived and interpreted in the light of experience and tradition. Analytical mathematics can be accommodated into this scheme by identifying program with proof. Within the methodology of conventional mathematics, an output is very often guessed or intuited and the program (proof) is sought. In computer graphics the output is self-vindicating.

Though I am arguing that the concept of mathematics should be broadened, one must of course draw boundaries somewhere. Does a toy kaleidoscope generate theorems of perceived type? Is, for example, a loaf of bread put out by an automated bakery and generated from raw materials by means of a recipe a theorem of the perceived type? Additional considerations will obviously enter and provide limitations.

Given the stochastic or fuzzy nature of the universe, with the possibilities of erroneous programs, erroneous execution, round-off error, etc., the theorems of the perceived type must be regarded as having validity only in a probabilistic sense. However, I believe that conventional "hand-crafted" theorems likewise have only probabilistic validity. This point of view was explained in some detail in Davis [1].

Acknowledgements. To Dr. K. Knowlton, Bell Telephone Laboratories, Murray Hill, New Jersey, for alerting me to certain mathematical possibilities in computer art and animation. To Professors U. Grenander, C. Strauss and P. Wegner for numerous discussions. To Professors S. Bergman, M. Kline, R. Vitale and R. B. Kelman for a number of trenchant comments.

Supplementary remarks.

§3. *The role of the visual image in mathematical discovery.* About the turn of the century, Poincaré divided mathematicians into two types: the geometers and the analysts. Geometers think about mathematical objects in pictures while analysts operate with formulas. Occasionally the same results have been obtained independently. Thus both Riemann (the geometer) and Weierstrass (the analyst) developed a theory of integrals of algebraic functions. In more recent times, the Bergman-Weil generalization of Cauchy's formula to several complex variables was probably developed by Bergman from geometric and by Weil from analytic considerations.

However, it appears to me that whatever the path taken in these investigations, the goal and the final formulation was essentially analytic. I look forward to a situation where the geometric element becomes more independent and marches less to the tune of the analytic.

René Thom [11] argues for the restoration of geometry from a pedagogical point of view. He puts forward the claim that "any question in algebra is either trivial or impossible to solve. By contrast the classic problems of geometry present a wide range of challenges."

The historical problem of the decline of the visual image in mathematics is one that is worthy of serious study. A mathematical Gibbon should undertake it. I do not believe it is a phenomenon limited to mathematics, but extends (even!) to the graphic arts. It is related to a general tendency of breaking up and recombination (e.g., cubist art) which emerged in the industrial age and has continued up through the current post-industrial age.

Discussions of this historical tendency with R. B. Kelman put him in mind of a pathological condition of dyslexia attendant upon some sorts of brain damage. This appears to be due in part to improper communication between brain areas. The spatial (geometric) functions may be performed in one area while the symbolic (algebraic) functions may be performed in another area. Within the mathematical culture we seem to be dealing with a dysfunction which has largely shut off the operation of the "geometrizing" area.

§6. *Relationship between computer language and visual effect.* This can be profound, as every language or collection of subroutines sets up limitations. *Example.* Two fonts of capital letters were created by two almost identical processes. In the first process, however, trigonometric interpolation was used while the second process used interpolation by cubic splines. The stylistic differences were sufficiently strong to be picked up by the eye. *Example.* A conventionally-trained artist produced a recognizable portrait of a colleague with a CALCOMP plotter using a certain subroutine that was available. What, I asked him, distinguished the result from a freehand drawing? He replied that the CALCOMP was producing strokes that were impossible by ordinary wrist and arm movements, so that the overall effect was different.

§7. Admittedly, the idea of the "theorem of perceived type" is somewhat vague and mysterious. Perhaps an analogy will dispel some of the fog by showing that the mysterious is, in fact, a commonplace experience within the psychology of perception.

The score of Mozart's Symphony No. 40 is a program. When the score is translated into sound by an orchestra playing in a standardized way it becomes the G Minor Symphony as commonly understood. The score and the music, though not physically identical, are aspects of the same thing. This symphony with its own musical themes, texture, tonalities, nuances, rhythms and patterns is unique. It is identifiable by many people. It is fairly stable (a few bad notes here and there will not make much difference), but nonetheless it is an aleatory process operating at a reasonably high probability level. It is capable of having judgments of various sorts applied to it. It is capable of having mathematical statements made about it, e. g., the average pitch is such and such, or certain parts are invariant under time translations. But it is self-vindicating in the sense that it needs no further intellectual amplification or retranslation into other non-aural modes in order to establish its integrity or to be appreciated. The G Minor Symphony represents a unique experience and, stretching a point, the passage from the score to the music might be said to constitute a "theorem of the perceived type."

It is interesting to note that the word "theorem" is derived from the verb "θεωρειν" which means "to look at."

The point of view advocated here is related to that developed by M. Polanyi in his book *Personal knowledge.*

§8. Apropos of the question of whether an automated loaf of bread or Mozart's Symphony No. 40 is a "theorem," in an essay written a number of years ago, James Bryant Conant once posed the problem of whether cooking is a branch of chemistry and, if it is, why is it not taught at Harvard. Conant's conclusion was that this is largely a matter of convention.

If one considers attempts to create computer music (admittedly not very successful, though Mozart himself was one of the first to write on the topic), then one may be more prone to accept the G Minor Symphony as defining a mathematical theorem or structure.

Probabilistic validity. The point made in Davis [1] is, briefly, that the verification of a mathematical proof requires examination of long symbol strings to see whether they follow the canons of mathematical deduction. As verification errors are inevitable, and are part of the real world, even within simple arithmetic, the theorems which emerge have only probabilistic validity. The longer the strings are the greater is the likelihood of error.

One sympathetic but traditional correspondent, reasserting the position of Platonic mathematics, writes:

> Absolute, universally accepted proof is an ideal and one which we may never attain. But ideals keep us striving in a definite direction. Justice is an ideal which is certainly not realized in our society but it does have value.

To this I add that prudent societies, while yearning for ideal justice, do well to provide themselves with courts of law which dispense pragmatic justice. It is therefore misleading to promulgate Platonism as the sole philosophy operative within mathematics. On this and on probabilistic validity see R. Thom [11].

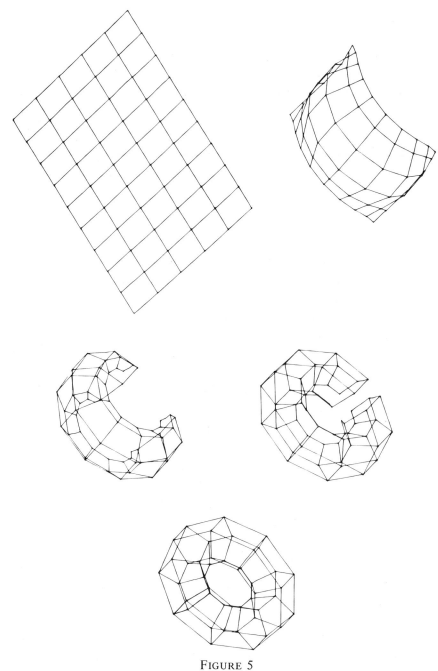

FIGURE 5

Stills from a computer-made movie: wrapping a rectangle to form a
torus. (Courtesy T. Banchoff and C. M. Strauss.)

BIBLIOGRAPHY

1. P. J. Davis, *Fidelity in mathematical discourse : Is one and one really two?*, Amer. Math. Monthly 79 (1972), 252–263.   MR 46 #3.

2. M. Kac, *Can one hear the shape of a drum?*, Amer. Math. Monthly 73 (1966), no. 4, part II, 1–23.   MR 34 #1121.

3. C. M. Strauss, *Computer-encouraged serendipity in pure mathematics.* Proc. IEEE 62 (1974).

4. T. Banchoff and C. M. Strauss, *On folding algebraic singularities in complex 2-space,* talk and movie presented at the AMS Meeting, Dallas, Texas, January 1973.

5. K. Knowlton, *The use of* FORTRAN*-coded* EXPLOR *for teaching computer graphics and computer art,* Proc. ACM AIGPLAN Sympos. on Two-Dimensional Man-Machine Communication, Los Alamos, New Mexico, 5–6 October 1972.

6. U. Grenander, *Computational probability and statistics,* SIAM Rev. 15 (1973), 134–192.

7. N. J. Nilsson, *Problem-solving methods in artificial intelligence,* McGraw-Hill, New York, 1971.

8. P. Wegner, *Three computer cultures : Computer technology, computer mathematics, and computer science,* Advances in Computers, vol. 10, Academic Press, New York, 1970.

9. Chandler Davis, *Materialist mathematics,* Festschrift for D. J. Struik (to appear).

10. M. Kline, *Logic versus pedagogy,* Amer. Math. Monthly 77 (1970), 264–282. MR 41 #3212.

11. René Thom, *Modern mathematics : An educational and philosophic error?,* American Scientist, Nov.–Dec. 1971, 695–699.

12. R. Arnheim, *Art and visual perception,* Univ. of Calif. Press, Berkeley, Calif., 1957.

13. ———, *Visual thinking,* Univ. of Calif. Press, Berkeley, Calif., 1969.

14. M. Polanyi, *Personal knowledge,* Univ. of Chicago Press, Chicago, Ill., 1958.

BROWN UNIVERSITY

Proceedings of Symposia in Applied Mathematics
Volume 20
1974

DUAL ORTHOGONAL SERIES:
A CASE STUDY OF THE INFLUENCE OF
COMPUTING UPON MATHEMATICAL THEORY

BY

ROBERT P. FEINERMAN, ROBERT B. KELMAN
AND CHESTER A. KOPER, JR.

1. Introduction. We present a brief description of a relatively new mathematical subject, dual orthogonal series, and of the decisive role computing has played in suggesting mathematical questions and their answers. As befits a case history, only a circumscribed set of our own investigations is discussed with just enough detail to outline the growth of ideas. For a more systematic presentation consult the References. We insert at this point an informal statement of the dual orthogonal series problem on the real line. Let c be a positive constant less than π, $\{a_n\}$ and $\{b_n\}$ sequences of positive constants, and $\{\psi_n(x)\}$ a complete orthonormal sequence in $L^2(0, \pi)$. Then given $f(x)$, find real coefficients $\{j_n\}$ such that

$$\sum_1^\infty j_n a_n \psi_n(x) = f(x), \qquad 0 < x < c,$$

$$\sum_1^\infty j_n b_n \psi_n(x) = f(x), \qquad c < x < \pi,$$

where equality may hold in an ordinary or generalized sense.

2. Origin of the problem. The study of orthogonal series arose, needless to say, in solving partial differential equations by separating variables [1]. For simplicity we restrict ourselves to Laplace's equation for which, in order to use the method, one must have "continuous" boundary conditions. Replacement

---

AMS (MOS) subject classifications (1970). Primary 42A60; Secondary 31A25, 35C10, 35J05, 40G05, 65N99.

by a simple kind of "discontinuous" boundary condition yields a dual orthog-onal series problem. For a rigorous description, see the Appendix in [2]. Here let us illustrate these notions by an example. Let h_1 and h_2 be nonnegative constants. We seek the potential $u(x, y)$ in $\{0 < x < \pi; y > 0\}$ with u bounded at infinity, $u_x = 0$ on $\{x = 0; y > 0\}$, $u = 0$ on $\{x = \pi; y > 0\}$, and with one of the three following boundary conditions holding:

(1) $$u_y = f(x) \quad \text{for } 0 < x < \pi, \, y = 0,$$

(2)
$$u_y = f(x) \quad \text{for } 0 < x < c, \, y = 0, \quad \text{and}$$
$$u = f(x) \quad \text{for } c < x < \pi, \, y = 0,$$

(3)
$$u_y = h_1 u + f(x) \quad \text{for } 0 < x < c, \, y = 0, \quad \text{and}$$
$$u_y = h_2 u + f(x) \quad \text{for } c < x < \pi, \, y = 0.$$

(*Notational aside.* A boundary condition of the form $u_n = -hu + f$, where n is the outward normal and h is a continuous positive function, is called a Newton boundary condition; the discontinuous boundary conditions (2) and (3) are often called "mixed" boundary conditions.) If we seek u as a separated variable solution it has the form, say,

$$u(x, y) = \sum_{1}^{\infty} \frac{j_n}{n - \frac{1}{2}} \cos\left(n - \frac{1}{2}\right)x \, \exp\left(-\left(n - \frac{1}{2}\right)y\right).$$

Imposing (1) we obtain in the usual Fourier fashion

(4) $$j_n = -\frac{2}{\pi} \int_0^\pi f(x) \cos\left(n - \frac{1}{2}\right)x \, dx.$$

If we impose (2) for the determination of j_n, we are led to the dual trigonomet-ric series equation

(5A) $$\sum_{1}^{\infty} j_n \cos\left(n - \frac{1}{2}\right)x = -f(x), \qquad 0 < x < c,$$

(5B) $$\sum_{1}^{\infty} \frac{j_n}{n - \frac{1}{2}} \cos\left(n - \frac{1}{2}\right)x = f(x), \qquad c < x < \pi,$$

whereas imposition of (3) yields

(6A) $$\sum_{1}^{\infty} j_n \frac{n - \frac{1}{2} + h_1}{n - \frac{1}{2}} \cos\left(n - \frac{1}{2}\right)x = -f(x), \qquad 0 < x < c,$$

$$(6B) \quad \sum_{1}^{\infty} j_n \frac{n - \frac{1}{2} + h_2}{n - \frac{1}{2}} \cos (n - \frac{1}{2})x = - f(x), \qquad c < x < \pi.$$

Thus, replacement of the continuous boundary condition (1) by either of the discontinuous boundary conditions (2) or (3) changes the determination of the Fourier coefficients from an orthogonal series problem into a dual orthogonal series problem.

Over the past twenty years a sizeable literature has grown up concerning dual orthogonal series deriving largely from applications in mechanical engineering [3]. The classical approach seeks closed form expressions for j_n analogous to (4); but in the case of dual series these take more complicated forms, e.g., multiple singular integrals [3]. This feature has precluded elaboration of a rigorous or general theory. Also, the classic method does not easily lend itself to dual series associated with mixed Neumann and Newton boundary conditions such as (6A) and (6B) important in heat transfer theory.

3. **Role of computing.** After many attempts at finding useful and general approximations for the above-mentioned singular integrals, we approximated $\{j_n\}$ by least squares. The simplicity and effectiveness of this procedure for dual equations leads us to ask in hindsight: Why was it not done before? Why did it take us so long to try it? The answer, in part, lies in our having been awed by the long string of successes [3] accruing to the classical analytic approach. Only under the impetus of concrete examples occurring in applications did we look elsewhere.

A few test runs established the efficacy of least squares approximations and work was started on program DUTSA (DUal Trigonometric Series Analyser) for the automated solution of dual trigonometric series [4]. Programming inexorably moved us to viewing the problem in a deeper and more abstract way. DUTSA could be transformed into DOSA (Dual Orthogonal Series Analyser) by changing the subroutine specifying trigonometric functions. This led to the definition of the dual orthogonal series problem given in §1.

However, for a while we were not able to find any general theorems to apply to this broad class of problems. At this point our interest in heat transfer theory helped. We carried out many calculations for dual trigonometric series associated with mixed Neumann and Newton boundary conditions [5]. From the computer output it was clear that convergence was faster for these series than for those dual series associated with a mixed Dirichlet condition. Referring to the literature ([6], [7]), we saw this reflected the fact that the solution to a two-dimensional potential problem at a point of discontinuity of a boundary

operator is better behaved if neither of the mixed conditions is a Dirichlet condition. With this clue we proved our first theorem (SIAM Rev. **15** (1973), 425).
It applied to dual trigonometric series of the type given by (6A) and (6B) and
depended delicately on inequalities of Schur [13] for bilinear forms for which
reason it was difficult to extend, e.g., to corresponding dual Bessel series. To
achieve the desired extension we generalized the problem to Hilbert space as follows (cf. [8]). Let **R** be a real abstract Hilbert space in which **P** and **Q** are
orthogonal complements. We denote by P and Q projection operators from
R onto **P** and **Q** respectively and by $\{\phi_n\}$ a complete orthonormal sequence.
The dual orthogonal series problem in **R** is: Given $g \in \mathbf{R}$, find a sequence $\{j_n\}$
in l^2 such that

$$\tag{7} \sum_{1}^{\infty} j_n(a_n P + b_n Q)\phi_n = g.$$

The least squares approximation problem of order N consists in finding constants, say $(j_1^N, j_2^N, \cdots, j_N^N)$, which minimize

$$\tag{8} I_N = \left\| \sum_{1}^{N} j_n(a_n P + b_n Q)\phi_n - g \right\|^2.$$

To see the connection between this formulation and (5A)–(5B) use the associa-
tions: **R** corresponds to $L^2(0, \pi)$; g corresponds to f; **P** is the set of all f in
$L^2(0, \pi)$ such that $f = 0$ (a.e.) for $c < x < \pi$; $a_n = 1$. From this the remaining
correspondences can be easily deduced. The set of linear algebraic equations in
j_n, namely $\partial I_N / \partial j_n = 0$, $n = 1, 2, \cdots, N$, formed the basis for the computations
in DUTSA.

It turns out that a_n/b_n tending to a positive limit as $n \to \infty$ corresponds
exactly to the relevant mixed boundary conditions being Neumann and Newton
conditions. This led to the following:

THEOREM. *If $\{a_n\}$ and $\{b_n\}$ are sequences of positive constants such that
$\{a_n\}$ is bounded above zero and a_n/b_n tends to a positive constant, then (7) has
a unique solution $\{j_n\}$ in l^2.*

This result is applicable to all dual Sturm-Liouville problems, including
multiple dual series, associated with mixed Neumann and Newton conditions [2].
Also it can be shown there exists $(j_1^N, j_2^N, \cdots, j_N^N)$ minimizing I_N and for
which $I_N \to 0$ under the hypothesis that $\{a_n\}$ and $\{b_n\}$ are merely positive
sequences [9].

The Hilbert space approach, however, is not subtle enough to provide an

existence theory for dual series in which one of the mixed conditions is a Dirichlet condition. Here computing has again provided a crucial insight. For many
such equations, e.g., (5A)–(5B) with $f \equiv 1$, we found poor pointwise numerical
convergence. Finally, we decided that maybe we did not have pointwise convergence. Thus motivated we have shown [10] for a class of dual trigonometric series typified by (5A)–(5B) that if f is of bounded variation the equations do
have a unique solution for which in general $j_n \neq o(1)$, a best possible growth
estimate for uniqueness is $j_n = o(n^{1/2})$ (cf. [11]) and the series in (5A) diverges
almost everywhere in the ordinary sense and converges almost everywhere in the
sense of Abel-Poisson. The proof depends upon conformal mapping and a strong
form of the two-dimensional reflection principle [12] so that it will be interesting to see if it can be extended to dual series other than trigonometric.

4. Closure. In summary, we have outlined for dual orthogonal series the
intimate connection between programming and computing on the one hand and
mathematical theory on the other—a connection which we believe will continue
in future investigations.

Finally, let us consider very briefly the following question: Mathematical
considerations aside, why would one bother with an analytic procedure for a
partial differential equation, e.g., separation of variables, rather than proceed
directly to a discretized solution? The answer depends on the specific application in mind. First, let it be noted that practically all methods require discretization at some point, e.g., the numerical evaluation of an integral as in (4). The
job of the applied mathematician is to discretize with discretion. If an answer is
to be computed only once, or if the geometry is highly irregular, then direct discretization procedures may prove superior or necessary. However, if answers must
be computed often, say more than a thousand times because of changes in design
parameters (which is likely if there are more than two parameters), then analytic
approximations may be advantageous because of their condensed form and shorter
computing times. It is intellectually easier to have a "feel" for an answer consisting of a series with a dozen terms than for a thousand arrays of three-dimensional vectors derived from a discretized solution. Further, data storage requirements are reduced for only the coefficients in these series need be stored as the
other factors in each term, e.g., trigonometric functions, Bessel functions, etc.,
are part of the compiler software or available in standard subroutine packages.

REFERENCES

1. J. Fourier, *The analytical theory of heat* (translation of 1822 edition), Dover, New
York, 1955.   MR 17, 698.

2. R. B. Kelman and R. P. Feinerman, *Dual orthogonal series*, SIAM J. Anal. Math. (to appear).

3. I. N. Sneddon, *Mixed boundary value problems in potential theory*, North-Holland, Amsterdam; Interscience, New York, 1966. MR **35** #6853.

4. R. B. Kelman and C. A. Koper, Jr., *Least squares approximations for dual trigonometric series*, Glasgow Math. J. **14** (1973), 111–119.

5. ———, *Separated variables solution for steady temperatures in rectangles with broken boundary conditions*, Trans. ASME Ser. C. J. Heat Transfer **95** (1973), 130–132.

6. N. M. Wigley, *Asymptotic expansions at a corner of solutions of mixed boundary value problems*, J. Math. Mech. **13** (1964), 549–576. MR **29** #2516.

7. ———, *Mixed boundary value problems in plane domains with corners*, Math. Z. **115** (1970), 33–52. MR **41** #7274.

8. B. P. Belinskiĭ, *Fourier series and integrals related to dual equations*, Zap. Naučn. Sem. Leningrad Otdel. Mat. Inst. Steklov. (LOMI) **15** (1969), 66–84. (Russian) MR **44** #748.

9. R. P. Feinerman and R. B. Kelman, *The convergence of least squares approximations for dual orthogonal series*, Glasgow Math. J. (to appear).

10. R. B. Kelman, *A Dirichlet-Jordan theorem for dual trigonometric series* (in preparation).

11. R. P. Srivastav, *Dual series relations. V. A generalized Schlömlich series and the uniqueness of the solution of dual equations involving trigonometric series*, Proc. Roy. Soc. Edinburgh Sect. A **66** (1963/64), 258–268. MR **30** #4113.

12. F. Wolf, *Extensions of analytic functions*, Duke Math. J. **14** (1947), 877–887.

13. I. Schur, *Bemerkungen zur Theorie der beschränkten Bilinearformen mit unendlich vielen Veränderlichen*, J. Reine Angew. Math. **140** (1911), 1–28.

H. H. LEHMAN COLLEGE (CUNY)

COLORADO STATE UNIVERSITY AND UNIVERSITY OF COLORADO MEDICAL CENTER

COLORADO STATE UNIVERSITY

Proceedings of Symposia in Applied Mathematics
Volume 20
1974

THE DESIGN AND USE OF AN UNDERGRADUATE
NUMERICAL ANALYSIS LABORATORY

BY

MYRON GINSBERG

Introduction. In most universities there is a substantial gap between the contents of the undergraduate level numerical analysis course and the introductory computer programming course. The former usually covers such topics as solution of equations, approximation, interpolation, as well as numerical differentiation and integration; often, however, no emphasis is placed on error analysis, algorithmic strengths and weaknesses, or on the factors affecting a method's computer implementation. In the programming course little or no attention is given to either a broad perspective of the language's structure or its limitations in scientific computation. Within an engineering curriculum it is vitally important to fill the void between these two courses in order to present a realistic view of the interaction of algorithms with their computational environment.

To meet the above-mentioned need, a numerical analysis laboratory is being designed in the Department of Computer Science and Operations Research in the Institute of Technology at Southern Methodist University. The lab's primary goal is to introduce an experimental approach which will promote understanding of the influences of software and hardware on reliability and efficiency of computerized numerical methods. It is hoped that the net effect of this project will be to help students (a) to gain appreciation of error propagation behavior; (b) to improve their programming ability for scientific problems; (c) to encourage individual experimentation with computational environmental factors; (d) to relate theoretical and experimental aspects of modern numerical mathematics.

This paper briefly outlines four proposed areas for inclusion in the lab:
(1) software influences; (2) hardware influences; (3) algorithmic characteristics;

AMS (MOS) subject classifications (1970). Primary 98A30, 98B15, 65-01, 68-01;
Secondary 65C99, 68A10.

and (4) error analysis. Alternatives for administrating the course as well as suggested support materials are also mentioned. This is a preliminary report of the lab's design; an expanded, more detailed discussion will be published by the author at a later date.

Software influences. The topics in the software section illustrate behavior of programs which the student might use directly or indirectly when implementing an algorithm. Experiments can be devised to respond to such questions as the following: How machine-and/or compiler-independent are higher-level languages used in scientific applications? What tests can be developed to exhibit software characteristics? When should variable precision arithmetic be utilized and what are its effects on program execution time and storage? What are the strengths and weaknesses of library routines? Compiler influences can also be included, such as optimized vs. nonoptimized code, rounding vs. truncation options, packing and conversion of data, or variations of numerical results and execution time with several compilers; some specific examples are given by Dorr and Moler [3] and Kahan [6]. Additional items which can be introduced by experiments, outside readings, or discussions are advantages and disadvantages of currently available symbolic algebraic systems, the quality of existing mathematical software, and the selection of a programming language for particular types of problems. The influence of any or all of the above factors on precision of computational results should be emphasized. The depth of material in this section of the course will depend upon the students' previous software experience and training.

Hardware influences. A major topic in the hardware area is the influence of various machine representations of floating-point numbers. Interesting observations can be made about the effects of word length, number base selection, normalization, rounding procedures, mantissa and exponent tradeoffs, as well as interaction amongst the above factors. Students should be made aware of the weaknesses of existing floating-point representations and possible differences between some of their properties and those of the real line; Cody [1], Kahan [6], and Forsythe [4] indicate that the associative, distributive, and closure laws can and have been violated on many computers. It is not necessary to provide much information about hardware structure to understand the conceptual difficulties alluded to above; rather it is best to exhibit a variety of behavior patterns and indicate various detection tests; for example, Malcolm [7] offers a program which permits the FORTRAN user to determine floating-point number base, number of bits in the mantissa, and whether a rounding or truncating option has been selected.

The potential effects of the new, large-scale parallel machines and microprogramming facilities should be discussed in the lab. Some of the design features of the CDC STAR, the TI ASC, and the ILLIAC IV can be presented along with their possible influences on error propagation and programming efficiency. Microprogramming options on some computers may offer assistance in numerical analysis problems. Shriver [9] outlines a few possibilities for using microprogramming to reduce computational overhead associated with several algorithms.

Numerous additional subjects can be included in the hardware portion of the course: uses of hybrid (analog and digital) computers in scientific computation; accuracy comparisons between mini-computers and large-scale, general-purpose machines; efficient use of hardware paging, auxiliary storage devices, and other computer resources. Problems in converting from base 10 representation (input) into machine representation and then converting internal results back into base 10 numbers (output) have been investigated by Matula [8]; some of his results can be verified by student experiments.

Algorithmic characteristics, This section of the lab is closely associated with the specific methods studied in the numerical analysis course. After exposure to several algorithms to solve a particular problem, the class can be asked to develop their own criteria for choosing the "best" algorithm; the selection process should use as inputs such factors as observed computer accuracy, theoretical error bounds, number and type of iterations, execution time, use of computational resources, and reliability of specific algorithmic implementations. Students can compare or complement their standards with some of the theoretical results from work done in computational complexity (e.g. see Traub [10]) and test the validity of these models in various situations. Experiments to determine optimal termination of a numerical technique can also be developed.

Several algorithmic properties should be observed to attempt to answer the following questions: Under what conditions does a method produce poor results? Can programming checks be defined to discern when these circumstances occur? What are the tradeoffs between parallel and serial techniques (in terms of efficiency, error propagation, etc.)? What algorithmic features influence the need for multiple or variable precision arithmetic? In partial response to the last question, one such common phenomenon is known as subtractive cancellation (a situation in which there is a subtraction between two nearly equal computer representations of floating-point numbers); both Forsythe [4] and Kahan [6] discuss this trait. It appears in many methods. Students should learn to recognize such situations and take appropriate computational measures to maintain a prespecified amount of accuracy.

Error analysis. No single, completely satisfactory technique is currently available to easily account for all errors (theoretical errors and all computer propagated errors) in all numerical analysis problems. Even if such a method did exist, it would most likely be too complicated to present at the undergraduate level. An experimental approach in this area allows students to observe and to monitor how various computational environmental factors discussed above affect results. Error bounding schemes (e.g. see Ginsberg [5]) can be introduced to check on the reliability of computer generated answers. Theoretical and experimental error bounds can be compared. Wilkinson's backward error analysis [11] can be applied to problems in linear algebra. For pedagological purposes a process graph approach (see Dorn and McCracken [2]) enables students to visually trace and observe effects of total propagated error through short computational sequences.

Alternatives for administering lab. There are several ways to incorporate into the curriculum the experimental approach suggested in this paper. A few alternatives are listed below: (a) integration into present numerical analysis courses with one additional hour per week; (b) a separate lab course (one to three credit hours) open to anyone who has taken introductory numerical analysis and computer programming courses; (c) seminar in mathematical software (for undergraduates and/or graduates); (d) a second course in scientific programming; (e) independent studies and/or special projects course; (f) self-instruction course at a computer terminal (i.e., a computer-assisted instruction approach). Any of the above choices or combinations of them could be chosen depending on the depth of coverage desired, student background, and personnel available. It can be administered by a computer science or mathematics department and/or taught via team teaching of persons from several departments.

It should be emphasized at this point that the proposed laboratory approach is intended to *supplement*, not replace, the contents of the traditional numerical analysis and programming courses. It may be possible to introduce a scaled down version of these topics within an existing course structure and minimize class time by use of extensive programming exercises, handouts, and outside readings. If this option is taken the instructor must carefully determine what subject matter, if any, would be displaced; under most circumstances the suggested material should not compose more than ten percent of total course content and should be well integrated.

Support material. Since many teachers may lack sufficient background in the proposed areas, it is imperative to provide several resources to implement the lab: a manual of experiments, a student workbook, and an extensive bibliography.

It is hoped eventually to produce a comprehensive set of student-tested experiments and make them available to interested groups. A student workbook or diary would be helpful in gathering programming data and observations; it could include graphs of numerous relationships and overviews of experimental results.

It is also necessary to have sufficient means to observe a variety of software and hardware effects. One way is to make students run all experiments on more than one computer and/or with several different compilers on a single machine. At many schools this will be adequate to produce a variety of numerical behavior. Another approach which is under investigation at Southern Methodist University is to create an arithmetic simulator with the capability of accepting a FORTRAN program and permitting the user to specify a floating-point representation of any currently available computer, or of his own design; then all the arithmetic in the program is performed with the selected representation. To provide such a flexible simulator and make it relatively machine independent for wide distribution to other universities is being considered if sufficient funding can be found. At the present time experiments are performed on several computers with several compilers.

Concluding remarks. This project was begun about nine months ago. Initial student reaction has been quite encouraging; most of the classes are composed of engineering and computer science majors. They accept the experimental approach and are eager to devise their own experiments to determine the influence of programming techniques on implemented algorithms. The students seem to be acquiring an intuitive feel for types of error propagation and an awareness of the multitude of factors affecting their computational results.

The author welcomes comments from interested individuals about the use of such an experimental approach to supplement a traditional undergraduate numerical analysis course.

BIBLIOGRAPHY

1. W. J. Cody, *The influence of machine design on numerical algorithms,* Proc. Spring Joint Computer Conference, AFIPS **30** (1967), 305–309.

2. W. S. Dorn and D. D. McCracken, *Numerical methods with FORTRAN IV case studies,* Wiley, New York, 1972.

3. F. W. Dorr and C. B. Moler, *Roundoff error on the CDC 6600/7600 computers,* SIGNUM Newletter **8** (1973), 24–26.

4. G. E. Forsythe, *Pitfalls in computation or why a math book isn't enough,* Amer. Math. Monthly **77** (1970), 931–956.

5. M. Ginsberg, *Introduction to the study of algorithms for computing upper and lower bounds to the exact solution of problems in numerical analysis,* Report No. CP-73024, Dept. of Computer Science and Operations Research, Southern Methodist University, Dallas, Texas, August 1973.

6. W. Kahan, *A survey of error analysis,* Proc. IFIP Congress 71, North-Holland, Amsterdam, 1972, pp. 1214—1239.

7. M. A. Malcolm, *Algorithms to reveal properties of floating-point arithmetic,* Comm. ACM **15** (1972), 949—951.

8. D. W. Matula, *In-and-out conversions,* Comm. ACM **11** (1968), 47—50. MR **39** #2360.

9. B. D. Shriver, Sr. *Microprogramming and numerical analysis,* IEEE Trans. Computers C-**20** (1971), 808—811.

10. J. F. Traub, *Computational complexity of iterative processes,* SIAM J. Comput. **1** (1972), 167—179.

11. J. H. Wilkinson, *Rounding errors in algebraic processes,* Prentice-Hall, Englewood Cliffs, N. J., 1963.

SOUTHERN METHODIST UNIVERSITY

Proceedings of Symposia in Applied Mathematics
Volume 20
1974

STATISTICAL AND NUMERICAL ANALYSIS:
A COMPUTER ORIENTED APPROACH

BY

ANDRE R. BROUSSEAU

This course was conceived and designed to attract students in the life sciences, government, business and economics who have had at least one semester of calculus. But it is equally useful for students in mathematics, applied mathematics and the physical sciences. The course is intended to provide students at the sophomore level with an introduction to the subject areas of statistics, numerical analysis and computer programming.

Several formats were suggested for implementation of this approach. But the following was chosen because a better blending of the three subject areas results than from any of the other formats. In general, this format is to introduce a specific statistical concept by presentation of a problem in which this concept is needed to solve the problem. This leads to an analysis of the numerical and/or combinatorial ideas relevent to the statistical concept. Various numerical techniques are discussed, and the most promising are then used in writing computer programs. Finally, the original statistical problem is solved using the computer program developed from the numerical methods which were utilized to solve the statistical concepts.

To illustrate this format, let us consider several examples. In each, an attempt is made to make the problem as realistic as possible.

A toll booth collector on a major toll road reports that 40% of the vehicles which pass his station are RV's. If 15 vehicles go through his toll station in the next 15 minutes, what is the probability that

(1) at least 10 of the 15 will be RV's,

(2) from 3 to 8 will be RV's.

Of course, any situation would do just as well. It is the success-failure outcome of the experiment that is of importance here. With the given problem as an

---

AMS (MOS) subject classifications (1970). Primary 98A25, 98A30.

introduction, the statistical concept of the binomial distribution is discussed. The binomial distribution is defined as:

If a binomial trial can result in a success with probability p and a failure with probability $q = 1 - p$, then the probability distribution of the binomial random variable x, the number of successes in n independent trials, is

$$b(x; n, p) = \binom{n}{x} p^x (1 - p)^{n-x}$$
$$x = 0, 1, 2, \cdots, n.$$
$$= \frac{n!}{x!(n - x)!} \; p^x (1 - p)^{n-x}$$

Once the statistical concepts involved have been discussed, the numerical questions inherent in the above equation are then considered. For example, what is "$n!$"? n-factorial, denoted $n!$, is defined to be the product $n(n - 1)(n - 2) \cdot \cdots 3 \cdot 2 \cdot 1$.

The next step in this development is that of writing a computer program which will utilize the mathematical expression of $n!$ and thus be able to calculate $b(x; n, p)$ for any x, n or p.

Note, however, that such a program only yields a specific value for a specified set of variables of the equation $b(x; n, p)$. To compute the values required to answer the given questions, it is necessary to derive the sum of the areas under the binomial curve. In this example, to answer the first question let X be the number of RV's which pass the toll booth. Then

$$P(X \geqslant 10) = 1 - P(X < 10) = 1 - \sum_{x=0}^{9} b(x; 15, 0.4).$$

This leads to a new numerical concept, that of summation, which is a fairly simple algebraic concept. That is, Sum $= \Sigma b_x$. The computer solution, however, requires both a special algorithm, namely $S = S + B(X)$, and the technique of LOOPing or iteration within the computer program.

The statistical problem may now be solved using the numerical techniques developed and the program written to generate answers to these techniques.

Consider a second example. An electrical firm manufactures light bulbs that have a length of life that is normally distributed with mean equal to 800 hours and a standard deviation of 40 hours. Find the probability that a bulb burns between 778 and 834 hours,

This problem appears to be similar to the first problem, the chief difference being that the data is no longer discrete. This problem deals with time and time is continuous. New techniques must be developed in all three areas. In statistics, the most common continuous distribution is that associated with the normal curve. By definition:

The density function of the normal random variable X, with mean μ and variance σ^2 is

$$n(x; \mu, \sigma) = \frac{1}{(2\pi)^{\frac{1}{2}} \sigma} \exp\left\{-\frac{1}{2}\left(\frac{x - \mu}{\sigma}\right)^2\right\}, \quad -\infty < x < \infty.$$

This equation is usually derived along with the various characteristics which hold for this curve. Some of the more tedious parts of the derivation are glossed over. But the purpose is to impress upon students the fact that each of the various density functions could be derived, that it is not simply an equation from above. To return to the problem at hand, the distribution of the useful life of these light bulbs is again given by an area, this time under the normal curve. The desired solution can be determined by

$$P(778 < X < 834) = \int_{778}^{834} n(x; \mu, \sigma)\, dx$$

$$= \frac{1}{(2\pi)^{\frac{1}{2}} 40} \int_{778}^{834} \exp\left\{-\frac{1}{2}\left(\frac{x - 800}{40}\right)^2\right\} dx.$$

Integration, of course, requires a slightly different approach than did the work with the computation of the area under the binomial curve. The process remains the same: It is a summation process. But integration is an infinite process and thus, the numerical techniques yield only an approximation. This means that the technique chosen must deal with accuracy of the solution desired.

Several numerical techniques are usually discussed. Among these are the trapezoidal rule and Simpson's rule. In addition, the speed of execution and the accuracy of each technique are also considered. Both of these methods suffer from excessive use of computer time with an additional fall-off in accuracy as successive iterations are accomplished. Thus, rather than use a counter to exit from the loop as was done in the program for the binomial distribution problem, the idea of a tolerance is introduced and used in this program. Once this has been accomplished, the program may be used to solve the given statistical problem using one of the numerical methods discussed.

A final example will serve to illustrate some additional points. Given a particular process in bio-chemistry, a study was made to determine the amount of converted sugar at various temperatures. We would like: (a) to determine what type of relationship exists between these two factors; (b) to be able to estimate the amount of converted sugar at some specified temperature; (c) to decide whether the model we are using is adequate.

By plotting the given data in a scatter diagram, it appears that the relationship is linear. The statistical concept will focus on the linear regression analysis. Each observation (x_j, y_j) in the sample satisfies the relation $y_j = a + bx_j + e_j$

where e_j is called the residual. Using a least squares method, an attempt is made to determine a and b such that the sum of the squares of the residuals is a minimum. This residual sum of squares, denoted SSE, is minimized:

$$SSE = \sum_{j=1}^{n} e_j^2 = \sum_{j=1}^{n} (y_j - a - bx_j)^2.$$

Differentiating SSE with respect to each of the variables a and b, and setting the partials equal to zero, the normal equations

$$na + b\sum x_j = \sum y_j, \qquad a\sum x_j + b\sum x_j^2 = \sum x_j y_j$$

are obtained.

This is simply a system of two equations in two unknowns with the added complication of the summations present. So the numerical technique of solving a system of m equations in n unknowns is discussed, using the Gauss elimination method with partial pivoting.

By this time in the course, students have experienced various types of summation processes. Class time is thus devoted to first explaining the Gauss elimination method. Once some problems are done using this method, then the partial pivot is introduced as a means of overcoming some of the problems encountered in running Gauss's method by itself.

This course has been extremely useful for students from all academic disciplines. The course provides each student with the necessary tools in numerical methods, statistics and computer methods which will be needed and used in the remainder of their academic careers. It has proven to be extremely useful for those students doing research on independent projects.

CENTER COLLEGE OF KENTUCKY

Proceedings of Symposia in Applied Mathematics
Volume 20
1974

SOME PROBLEMS IN COMPUTATIONAL PROBABILITY [1]

BY

MARCEL F. NEUTS

ABSTRACT. Recent work in the algorithmic solution of problems in stochastic models, by the author and several collaborators, is surveyed. The educational value of computational probability as a complement and extension of the usual approaches is also stressed.

1. Introduction. During 1971, an intensive effort in developing algorithmic solutions to problems in probability theory and stochastic models was initiated by the author at Purdue University. This work benefited greatly from the interest and collaboration of Professor Eugene M. Klimko, who is now at the State University of New York at Binghamton, New York, and of Mr. David Heimann and Mr. William D'Avanzo, graduate students at Purdue.

There are two main reasons for our continued interest in this area. *In research* in probability models, notably in the theory of queues, we were struck by the almost total absence of algorithms to implement the results of the rich mathematical structure of such models. Only a small number of explicit formulas are available, and even fewer among those lend themselves well to numerical computation. The practioner will therefore often ignore the theoretical results altogether and resort instead to Monte Carlo simulation, although the latter is time consuming and does not offer a high degree of accuracy. We found that a careful analysis of the structure of probability models frequently leads to efficient algorithms to produce the answers to questions, which otherwise would be analytically untractable.

In the teaching of basic courses in probability and stochastic processes, there is an exclusive concern with the rare problems which have simple explicit solutions

AMS (MOS) subject classifications (1970). Primary 60–02.

[1]Research sponsored by the Air Force Office of Scientific Research, Air Force Systems Command, USAF, under grant AFOSR-72-2331. The United States Government is authorized to reproduce and distribute reprints for governmental purposes notwithstanding any copyright notation hereon.

or an easy limit behavior. By an undue stress on elegance, most presentations
never teach the student to handle the complexity found in most real problems.
Through the algorithmic approach the student gains a deeper insight in the probabilistic reasoning, involved in the solution of more complex problems. In developing computer programs for two or three problems per semester, he learns to pay
more attention to details and particular cases than is required for most textbook
problems.

Detailed discussions of the algorithmic solution to several problems in
queueing theory have appeared to date in a series of four papers in *Naval Research
Logistics Quarterly*. Several additional papers are in preparation. The pedagogical material, much of which is novel in form and content, will be presented in a
book by Professor Klimko and the author.

2. Algorithmic research in stochastic models. The stochastic models, which
we have studied to date, are related to the single server queue and to the machine
repair problem. These models are structurally well understood, but very few
usable analytic results are available. A presentation of the analysis of such models
is infeasible here. Instead we shall summarize our experience in several general
observations and mention a number of further research problems, which have
been suggested by the algorithmic solution.

A. *The importance of discrete parameter models.* Problems in stochastic
processes typically involve the use of complicated multivariate or conditional distributions. In continuous time or with a continuous state space, the direct
numerical analysis of even simple models may involve unmanageable storage problems or millions of costly numerical integrations. Such models often arise from
discrete phenomena, which for reasons of tradition or in the hope of approximation results are formulated in terms of continuous variables. Whether this is the
case or not, there is frequently a discrete formulation of the model, which is
computationally tractable and offers a high degree of insight into the phenomenon.

The time has come to reexamine a number of classical mathematical models
and to distinguish between essential qualitative assumptions on the one hand,
and properties which are postulated in the hope of deriving an explicit or an
asymptotic solution. In the latter category, we found that the assumptions which
lead to analytic tractability are rarely the same as those which make computation
feasible and economically useful. The outstanding example is the use of discrete
parameter models. We were able to study the transient behavior of long waiting
lines for a certain discrete time version of a queueing problem, while the usual
continuous time model remains prohibitively complicated and untractable.

Other classical assumptions of stochastic models may be critically examined from the same viewpoint. The analysis of an unbounded queue is usually *conceptually* easier than that of a bounded queue. The assumption of an unlimited waiting capacity is very useful in a theoretical analysis, but it can be best abandoned in most algorithmic solutions. This should certainly be done where the traditional assumptions become a source of purely academic problems of little or no practical importance.

B. *The importance of recursive methods.* The majority of formulas in terms of integral transforms which abound in the theory of stochastic models are computationally useless. It is far more promising in most cases to return to the recurrence relations, which are hidden away in them [4]. For the same model, there are usually several mathematically equivalent sets of recurrence relations. These may however be vastly different in computational usefulness, depending on such things as the order of the recursion, the availability of intermediate quantities which can be efficiently computed, the amount of memory storage and the processing time required. There is usually a trade-off between storage requirements and processing time. He who implements an algorithm should have as many alternate computational approaches available as possible so that he may find one that is compatible with the limitations of the computer and the economics of processing time. This is a nontrivial "art," learned only by experience a and insight. It is the prime reason for a probabilist to become involved with algorithmic analysis. His insight into the structure and qualitative behavior of the model is usually far more important than are classical techniques of computer programming or numerical methods in work of this nature.

C. *Limit theorems versus approximation theorems.* Much of the value of limit theorems in applied mathematics lies in their use as approximation theorems. In stochastic models, many important limit theorems are known, but only in rare cases do we have sharp results on the rate of approach to the limit. In really complicated situations, numerical exploration may be the only way of gaining information on this rate, and therefore on the quality of the limit theorem for approximation purposes.

In queueing theory, for example, the practical worker is often only conversant with so-called steady state results. These do not convey any information on the growth of practical queues which are unstable, but are considered only over a finite interval of time. This is usually well understood, but there is a much more serious trap which is not so obvious, even in the case of stable queues. The equations for the steady state of queues are usually derived from the ergodic theorem, in some manner. The ergodic theorem "averages out" all oscillatory behavior. In

[5], a numerical example is given of a very stable queue whose length over time exhibits strong oscillations. For this queue, the steady state probability distributions are easy to compute, but could lead to disastrous practical conclusions unless carefully interpreted in the light of the time-dependent behavior.

This example suggests that suitable theoretical measures of oscillatory behavior for the path functions of the classical stochastic models are needed. Even when these become available, it will remain useful to have illustrative examples to show the dangers inherent in the uncritical use of limit results.

We shall now discuss some results and problems which arose from our work in stochastic models, but which appear to be of wider interest and applicability.

D. *Fáa di Bruno's formula.* In the study of the higher moments of the busy period of a single server queue [3], we need to evaluate higher order derivatives, at $x = 0$, of a function which is the composition of three (or four) functions with known derivatives. Analytic differentiation soon becomes too unwieldy to remain practical. Instead the classical formula of Fáa di Bruno,

$$\frac{d^n}{dx^n} f[g(x)] = \sum_{r=1}^{n} \left[\frac{d^r f(y)}{dy^r}\right]_{y=g(x)} \sum_{E_{n,r}} \frac{n!}{j_1! \cdots j_n!} \left[\frac{g^{(1)}(x)}{1!}\right]^{j_1} \cdots \left[\frac{g^{(n)}(x)}{n!}\right]^{j_n},$$

was computationally implemented. The second summation is over the set $E_{n,r}$ of n-tuples $(j_1, \cdots, j_n)$ of nonnegative integers, satisfying

$$j_1 + j_2 + \cdots + j_n = r, \qquad j_1 + 2j_2 + \cdots + nj_n = n.$$

Professor Klimko was able to generate all such n-tuples for n up to *fifty* in an elegant, systematic way [2]. For $n = 50$, the number of such n-tuples is on the order of 204,000. The Fáa di Bruno indices were stored on a magnetic tape and a very compact algorithm was programmed to evaluate derivatives of order up to fifty by using this classical differentiation formula.

Beyond order fifty the processing time of the algorithm becomes prohibitively long. Because of the enormous number of operations involved, we were initially very concerned about the numerical accuracy of the final results. Fortunately the desired moments were analytically tractable in a number of special cases, which retain all the numerical qualities of the general case. By comparison we were able to demonstrate that, at least for the queueing application, our algorithm exhibits a very high degree of numerical stability. The numerical results agreed to eleven significant digits at least, with the special cases, using only single precision on the CDC 6500 computer at Purdue. This is an excellent instance of a classical, but hitherto quite useless formula, acquiring definite practical value in the computer age. The details of its implementation require sophisticated computer techniques but should prove to be useful in other problems of applied mathematics.

E. *A class of integral equations.* Equations of the type

$$F(x) = \sum_{n=0}^{\infty} A_n(\cdot) * F^{(n)}(x), \quad \text{for } x \geqslant 0,$$

where $F(\cdot)$ is an unknown probability distribution on $[0, \infty)$, whose n-fold convolution is denoted by $F^{(n)}(\cdot)$, occur frequently in stochastic models. The coefficient functions $A_n(\cdot)$ are probability mass functions on $[0, \infty)$; their sum $A(\cdot) = \sum_{n=0}^{\infty} A_n(\cdot)$ is a probability distribution on $[0, \infty)$.

The iterative solution of the analogous nonlinear difference equation is discussed in [1]. There are several interesting numerical problems associated with this equation. The first concerns the errors resulting from truncation both in n and in x of the summation appearing in the right-hand side. The second problem is to evolve an *efficient* algorithm to evaluate convolution polynomials of the form $\sum_{n=0}^{N} a_n * f^{(n)}$, even for the case where a_n and f are sequences. We used a convolution analogue of Horner's algorithm for ordinary polynomials, but even this method soon becomes prohibitive. The third question is related to determining *a priori* a constant T such that $F(T) > 1 - \epsilon$. For every $\epsilon > 0$, T should be as close as possible to the $(1 - \epsilon)$-quantile of the distribution $F(\cdot)$. If such a constant T is known, the equation may be solved recursively on the interval $[0, T]$. Since the computation time of a recursive solution depends quadratically on T, a close value of T is highly desirable. We do not have complete and satisfactory solutions to these problems, but some interesting side results were obtained. It was found, for instance, that the upper bound on the $(1 - \epsilon)$-quantile *derived from Markov's inequality* is a strictly convex function of the order of the moment of $F(\cdot)$ which is used in that inequality [6].

3. **Teaching computational probability.** We have not been concerned with computer-aided instruction in elementary probability or statistics courses. In such courses the computer is used primarily for illustrative purposes and there is rarely a need for detailed analysis and planning of the algorithms. Our aim has been instead to teach the more advanced student to integrate the algorithmic approach as early and as fully as possible into the solution of more complex problems. In the preparation of examples and problems, we found that it is usually necessary and desirable to go far beyond the standard text book problems in probability. Our audience of graduate students in statistics and computer

science felt properly challenged and learned not only the development of algorithmics, but also a substantial amount of probability in the process.

The following is a partial list of topics we selected for the course and for detailed discussion in a text book, which is in preparation:

1. A large variety of problems related to order statistics, nonclassical probability distributions, geometric probability models, first passage times and combinatorics, all of which are amenable to solution by well-chosen recurrence relations.

2. A discussion of root-finding techniques and of numerical integration as they apply to problems arising in probability and statistics.

3. Illustration of the inclusion-exclusion formula in fairly large combinatorial problems, with a discussion of the numerical difficulties, associated with this formula.

4. A systematic treatment of finite Markov chains and their use in queues, machine repair models and inventories.

5. Erlang's method of phases and its implementation in the numerical solution of several stochastic models.

6. A discussion of the basic techniques of Monte Carlo simulation.

7. A discussion of the joint probabilities for the point counts in the four hands in the game of bridge. This is a very instructive case study of the prior analysis required in a complex problem *before* it is submitted to the computer.

A course in computational probability can only be successful if very active student participation is elicited. We therefore prepared an extensive collection of problems, most of which were novel or related to the specialized literature. Shorter problems are assigned to one or more of the students and require only a small amount of work. In addition to this, each student is to select a project-problem to be completed by the end of the semester. Examples of such projects are the systematic investigation of the probability distributions associated with (say) a statistical procedure for selection and ranking, the study of a simple stochastic model or the computation of shortest confidence intervals for a given class of probability distributions.

We shall conclude with a brief discussion of the relative weights which, in our opinion, should be given to probability theory, numerical analysis and computer programming in a course in computational probability. Ideally, students should come to such a course with a good background in all three of these areas. In reality, they rarely do. Our approach has been to emphasize the probabilistic arguments strongly, but to call freely on known theorems of numerical analysis, which are presented without proofs.

Exact mathematical analysis of the accuracy of algorithms is usually difficult. Unless the student is unusually competent in numerical analysis, he will only rarely perform such analysis on concrete problems. Instead, we teach in most cases to look for test examples which exhibit known properties. If at all possible, we suggest examples which have symmetry properties or a known asymptotic behavior to test the accuracy and the correctness of the proposed numerical method. For most probability problems such test examples are readily available.

Computer programming is treated as a tool in, and not a purpose of, this course. In our experience, students who see clever solutions to complex problems also write efficient codes. In the latter part of the course, we discuss a number of problems in which the method of solution is dictated by the need to work within the limitations of the computer's memory size. In order to prepare for these problems, we stress efficient computer use in a number of much simpler examples.

Our experience is too limited to assess the value of a course in computational probability in the graduate training of a statistician or probabilist. The students we had seem to have become much better problem solvers—what more can one ask for?

Remark. The following references are directly related to the content of the paper. A more extensive list of references to other work may be found in T. P. Bagchi and J. G. C. Templeton, *Numerical methods in Markov chains and bulk queues*, Springer Lecture Notes in Economics and Mathematical Systems, No. 72, Springer-Verlag, Berlin and New York, 1972.

REFERENCES

1. David Heimann and Marcel F. Neuts, *The single server queue in discrete time—Numerical analysis.* IV, Naval Res. Log. Quarterly **20** (1973), 753–766.

2. Eugene Klimko, *An algorithm for calculating indices in Faà di Bruno's formula*, Nordisk Tidskr. Informations-Behandling (BIT) **13** (1973), 38–49.

3. Eugene Klimko and Marcel F. Neuts, *The single server queue in discrete time—Numerical analysis.* II, Naval Res. Log. Quarterly **20** (1973), 305–319.

4. Marcel F. Neuts, *The single server queue in discrete time—Numerical analysis.* I, Naval Res. Log. Quarterly **20** (1973), 297–304.

5. Marcel F. Neuts and Eugene Klimko, *The single server queue in discrete time—Numerical analysis.* III, Naval Res. Log. Quarterly **20** (1973), 557–567.

6. Marcel F. Neuts and David B. Wolfson, *Convexity of the bounds induced by Markov's inequality*, J. Stoch. Proc. and Appl. **1** (1973), 145–149.

PURDUE UNIVERSITY

Proceedings of Symposia in Applied Mathematics
Volume 20
1974

THE INFLUENCE OF COMPUTING ON GENERALIZED INVERSE APPLICATIONS IN STATISTICAL ANALYSIS

BY

CECIL R. HALLUM

Introduction. In 1920 E. H. Moore [8] established the existence and uniqueness of a "general reciprocal" matrix associated with any rectangular matrix with complex elements but the material received little attention until the concept was independently rediscovered by Penrose in 1955 [9]. Later, Rao [10] showed that a general inverse, called the generalized inverse, with a much weaker definition than that of Moore and Penrose is sufficient in dealing with problems of linear equations. Such matrices received increasing attention in applications primarily because of the fact that the general solution to the consistent equations $Ax = b$ is given by

$$x = A^g b + (I - A^g A)z$$

where A^g is the generalized inverse and z is arbitrary (but conformable), but also because of the advent of better and more sophisticated methods and means of computing. In particular, since the Penrose rediscovery, statistical analysis has undergone a great unification, particularly so in the areas of estimation and testing in the linear statistical model. Personally, it has been my experience as a teacher, student, and researcher that the generalized inverse has provided an ease, never before realized, in understanding, remembering, and explaining the classical, as well as many recent, results in statistical analysis. The concept of generalized matrix inversion with the aid of an electronic computer continues to play an increasingly important role in the formulation and solution of many complex problems, not only for the statistician but for mathematicians, engineers, and others as well. I think most statisticians would agree that the computer has been powerfully influential (if not indispensible) in at least the areas of raw data analysis, the teaching of statistics, and in methodological statistical research. In this paper,

AMS (MOS) subject classifications (1970). Primary 62J10.

a rough overview of some of the influences of computing on generalized inverse applications in statistical analysis is presented along with a specific case in which a direct influence has recently reached the classroom. In particular, the analyses in fixed, mixed, and random effects models have been simplified via the coalition of generalized inverse theory with new and better computing methods. One cannot hope, in so short a time, to do justice to the many papers, ideas, and techniques that have contributed to the young, but already extremely large, field of generalized inverses and their applications. Sources for witnessing many of the influences of computing on generalized inverse applications in statistical analysis may be observed by referring to the extensive bibliographies, for example, included in the monographs of Boullion and Odell [1] and Rao and Mitra [12].

A categorization of, and comments on, a few computational schemes. The defining properties of the generalized inverse are given as follows: Given any $m \times n$ matrix A, if we denote the Rao generalized inverse by A^g, then the defining equation for A^g is $AA^gA = A$ while for the Moore-Penrose pseudo-inverse, denoted by A^+ (and which is unique for each A), the following must hold: (1) $AA^+A = A$, (2) $A^+AA^+ = A^+$, (3) $(AA^+)^T = AA^+$, and (4) $(A^+A)^T = A^+A$.

To give a rough categorization of some (this list is certainly not exhaustive) of the existent computational schemes, the criterion used was primarily how well these techniques lend themselves to a computer programmable solution which is dependent upon (1) efficiency considerations (e.g., accuracy), (2) storage considerations, and (3) execution time. These categorizations are influenced particularly by comments and conclusions from Dekerlegand [3] and Rao and Mitra [12].

CATEGORIZATION OF SCHEMES

Suitable for
Numerical Calculations

1. Penrose Method II
2. Greville Method
3. Rust, Burrus and Schneeberger (R. B. S.) Method
4. Ben-Israel Method
5. Pyle Method

Useful in
Theoretical Investigations

1. Generalized inversion schemes based on matrix factorization:
   (a) Rank factorization (Householder [7])
   (b) Diagonal factorization (Rao [11])
   (c) Hermite canonical form
2. Penrose Method I
3. Hestenes Method

Other methods continue to appear in the literature and a basic influence of computing that continues to influence generalized inverse applications is that of computationally evaluating new schemes and comparing them with others to justify the use of better methods of generalized inversions. From Dekerlegand's results, the conclusion is that although the R.B.S. scheme is close in comparison, the method of Greville takes precedence as a computational method especially suited for the computer. In particular, (1) Greville's method does not require the inverse of any submatrix, (2) it does not require the factorization of any matrix, (3) it does not require considerable matrix manipulations which involve matrix multiplication. For future reference, let us briefly review Greville's method and that of Penrose (Penrose Method II).

Greville Method. Let $A = (A_{k-1}, a_k)$ where a_k is the kth column of A and A_{k-1} is the submatrix of A consisting of the first $k-1$ columns, and compute $d_k = A_{k-1}^+ a_k$ and $c_k = a_k - A_{k-1}d_k$. If $c_k \neq 0$, let $b_k = c_k^+$. If $c_k = 0$, compute $b_k = (1 + d_k^T d_k)^{-1} d_k^T A_{k-1}^+$. Then

$$A^+ = A_k^+ = \begin{pmatrix} A_{k-1}^+ - d_k b_k \\ b_k \end{pmatrix}.$$

To initiate the process, take $A_1^+ = 0$ if a_1 is a zero vector; otherwise, $A_1^+ = (a_1^T a_1)^{-1} a_1^T$.

Penrose Method II. Let $B = A^T A$ and define a sequence of matrices C_j, $j = 1, 2, \cdots, r$, by

$$C_1 = I, \qquad C_{j+1} = (1/j)I \, \mathrm{Tr} \, (C_j B) - C_j B$$

where "Tr" denotes the trace operator. It can be shown that $C_{r+1}B = 0$ and $\mathrm{Tr}(C_r B) \neq 0$, where r is the rank of B, and

$$A^+ = r C_r A^T / \mathrm{Tr}(C_r B).$$

Two additional identities which frequently prove useful in computing A^+ are (1) $A^+ = (A^T A)^+ A^T$ and (2) $A^+ = A^T (A A^T)^+$, the one selected being determined by which matrix, i.e., $A^T A$ or $A A^T$, has the smaller dimensions (e.g., for the pseudo-inverse of an $n \times 1$ column vector a, $a^T a$ would be preferable over $a a^T$ since the former is a scalar, the latter an $n \times n$ matrix). Incidently, the primary difficulty that has been experienced on the computer for the Penrose Method II is that, for large matrices, $C_{r+1}B$ tends to converge away from the zero matrix.

Influence on teaching. If we can teach an entering freshman to obtain the inverse of a nonsingular matrix using any of the standard procedures (which we do—or I should say, attempt to do—in most basic introductory mathematics courses) then it certainly would not be any more difficult (in fact, I believe less difficult) to teach that individual to calculate the Moore-Penrose pseudo-inverse using either the Penrose Method II or the Greville Method as previously outlined. The reasons for this conclusion have been stated already in regard to the Greville Method. For many of the same reasons, the Penrose Method is especially suited for teaching purposes as well: Only the basic operations of scalar multiplication, matrix addition, matrix multiplication, and the trace are required, all of which, with the exception maybe of the trace, are taught in the standard basic freshman mathematics course. The point I wish particularly to stress is that the time is long overdue that we teach the Moore-Penrose pseudo-inverse in place of the standard computational schemes presently taught for the inverse. With this, we can teach at the introductory level that every candidate solution to the consistent set of equations $Ax = b$ is given explicitly and very compactly by

$$x = A^+ b + (I - A^+ A)z$$

where z is any arbitrary vector (the only restriction being, of course, that it be conformable for multiplication). Some advantages of this approach are: (1) The computational schemes for the pseudo-inverse are just as teachable (in fact, I believe more so than, for example, the cofactor-determinant procedure or the elementary row operation procedure) as what is currently taught. (2) If A should be nonsingular, the standard solution $x = A^{-1}b$ results anyway. (3) The above expression for the general solution for x demonstrates to the student the possibility of many solutions, namely one for each arbitrary choice of z. (4) A necessary and sufficient condition for mathematical consistency of $Ax = b$ is $AA^+ b = b$, which provides a check for this situation (see Graybill [4]). (5) Even when $Ax = b$ is not consistent, one can teach the fact that $x = A^+ b$ provides a least squares solution. (6) This approach would provide students with a sufficient background for subsequent applications of the generalized inverse to analyses in analysis of variance, regression analysis, and analysis of covariance models which may be presented very efficiently at the introductory (noncalculus prerequisite) level.

Elaborating further in regard to (6), it is well known that the designs in analysis of variance, regression analysis, and analysis of covariance are completely characterized by the general linear model $Y = X\beta + \epsilon$ (along with the standard assumptions $E(\epsilon) = 0$, $E(\epsilon\epsilon^T) = \sigma^2 I$ on ϵ), with β possibly restricted to satisfy the constraint $R\beta = t$ (which may be assumed or inherent in the system), and with the standard hypothesis for test purposes being $H_0: \Lambda\beta = h$. In particular, all such designs are completely specified by the known quantities: the vector of observations Y, the design matrix X, the restriction matrix R, the vector t, the hypothesis matrix Λ, and the vector h. Moreover, explicit expressions for the form of the estimate $\hat{\beta}$ from the class of linear estimates of minimum bias norm (the norm being $\|E(\hat{\beta}) - \beta\| = [(E(\hat{\beta}) - \beta)^T(E(\hat{\beta}) - \beta)]^{1/2}$) which minimizes the risk matrix $E[(\hat{\beta} - \beta)(\hat{\beta} - \beta)^T]$ is given by

$$\hat{\beta} = R^+t + H^+(h - \Lambda R^+t) + M^+[Y - XR^+t - XH^+(h - \Lambda R^+t)]$$

where $H = \Lambda(I - R^+R)$ and $M = X(I - R^+R - H^+H)$ (see Hallum, Lewis, and Boullion [6]); if $t = 0$ and $h = 0$, as is the case in the standard designs, this simplifies very nicely to $\hat{\beta} = M^+Y$. Moreover, the standard test statistic for testing $H_0: \Lambda\beta = h$ has the general form

$$F = \frac{k}{r} \frac{Z^T(XC^+X^T - MM^+)Z}{(Y - XR^+t)^T(I - XC^+X^T)(Y - XR^+t)}$$

where

$$Z = Y - XR^+t - XH^+(h - \Lambda R^+t), \qquad C = (I - R^+R)X^TX(I - R^+R),$$

$$r = \text{Tr}(XC^+X^T - MM^+) \qquad \text{and} \qquad k = \text{Tr}(I - XC^+X^T).$$

Again in the standard designs $t = 0$ and $h = 0$, in which case

$$F = \frac{k}{r} \frac{Y^T(XC^+X^T - MM^+)Y}{Y^T(I - XC^+X^T)Y}.$$

The distribution of F under H_0 is, of course, the central F-distribution with degrees of freedom r and k, respectively, with the hypothesis being rejected at the α-significance level provided $F > F_{1-\alpha}(r, k)$. The standard ANOVA table has a convenient expression as well (see Hallum, Boullion, and Odell [5]); a few advantages that may be immediately noted are: (1) A very compact program can handle these results. (2) For the benefit of the student the analysis requires one to remember a single F-statistic and, at most, a knowledge of matrix multiplication and the trace operator. (3) The problem of having to estimate missing observations in order to achieve a balanced design is eliminated. (4) The analysis does not require matrix rank assumptions of any kind. The primary disadvantage is that of a possible lack of computer storage and, of course, inherent round-off errors. However there remains potential to relax this problem considerably by utilizing some of the recent computing ideas such as: (1) Use the method of linked storage detailed in Tewarson [13] to store and manipulate with the analysis of variance part of the design matrix, which is simply that part of the partitioned design matrix having all zeros and ones (predominantly zeros). (2) Partition X and use Cline's [2] result for the pseudo-inverse of a partitioned matrix.

In concluding the discussion of the influence on teaching, let me emphasize that the above approach was utilized in the classroom by a cross-section of students from the separate disciplines, including sophomores through seniors (most having very little mathematical background). Approximately three lectures were used to cover the necessary matrix manipulations required. The results were very encouraging: (1) We were able to cover several more designs than can ordinarily be covered in the same amount of time. (2) The students responded quite favorably to the simplicity of construction of the design matrix and the hypothesis matrix. (3) Not only was the material easier for them to understand, but by taking this approach, there was additional time to spend on the concept of designing experiments.

Conclusions. With the computing facilities and methods that exist today, it is clear that applications requiring the generalized inversion of matrices can be faced. Much work does lie ahead, but the results evidenced already are certainly very promising indeed. They would not have been possible had it not been for the influence of computing on generalized inverse applications in statistical analysis.

REFERENCES

1. T. L. Boullion and P. L. Odell, *Generalized inverse matrices*, Wiley, New York, 1971.

2. R. E. Cline, *Representations for the generalized inverse of a partitioned matrix*, J. Soc. Indust. Appl. Math. **12** (1964), 588–600. MR **30** #3106.

3. R. J. Dekerlegand, *Analysis of generalized inverse computational schemes*, Master's Thesis, University of Southwestern Louisiana, Lafayette, Louisiana, 1967.

4. F. A. Graybill, *Introduction to matrices with applications in statistics*, Wadsworth, Belmont, Calif., 1969. MR **40** #2688.

5. C. R. Hallum, T. L. Boullion and P. L. Odell, *Parameter estimation and hypothesis testing in the restricted general linear model*, J. Indust. Math. **23** (1973), 1–25.

6. C. R. Hallum, T. O. Lewis and T. L. Boullion, *Estimation in the restricted general linear model with a positive semidefinite covariance matrix*, Communications in Statistics **1** (1973), 157–166.

7. A. S. Householder, *The theory of matrices in numerical analysis*, Blasidell, New York, 1964. MR **30** #5475.

8. E. H. Moore, *On the reciprocal of the general algebraic matrix*, Bull. Amer. Math. Soc. **26** (1920), 394–395. (Abstract).

9. R. Penrose, *A generalized inverse for matrices*, Proc. Cambridge Philos. Soc. **51** (1955), 406–413.   MR **16**, 1082.

10. C. R. Rao, *A note on a generalized inverse of a matrix with applications to problems in mathematical statistics*, J. Roy. Statist. Soc. Ser. B **24** (1962), 152–158. MR **25** #1596.

11. ———, *Linear statistical inference and its applications*, Wiley, New York, 1965. MR **36** #4668.

12. C. R. Rao and S. K. Mitra, *Generalized inverse of matrices and its applications*, Wiley, New York, 1971.

13. B. Tewarson, *Sparse matrices*, Blasidell, New York, 1973.

LOYOLA UNIVERSITY

Proceedings of Symposia in Applied Mathematics
Volume 20
1974

ON USING THE ELECTRONIC ANALOG COMPUTER
TO ILLUSTRATE MATHEMATICAL CONCEPTS

BY

TYRE A. NEWTON

1. Introduction. Perhaps the mathematical community is not as aware as it might be of the electronic analog computer which processes information in the form of voltages, as an alternative to the digital computer which processes information in the discrete form. In fact, the analog computer is well suited for illustrating in a dramatic manner the kinematic and geometric aspects of mathematics, thereby supplementing the deductive, epsilon-delta viewpoint of which has been so heavily emphasized in recent decades.

2. The basic analog computer—a mathematician's view. To the mathematicians, the basic analog computer, such as the one shown in Figure 1, is essentially an electronic realization of the problem

(2.1) $$dx/dt = \mathbf{F}(t, \mathbf{x}) \quad \text{for} \quad t \geqslant 0,$$

$$\mathbf{x}(0) = \mathbf{x}_0,$$

where the boldfaced letters represent vector quantities, or, equivalently,

(2.2) $$\mathbf{x}(t) = \mathbf{x}_0 + \int_0^t \mathbf{F}(\tau, \mathbf{x})d\tau \quad \text{for} \quad t \geqslant 0.$$

The section (b) of the analog computer shown in Figure 1 is wired to realize electronically the function $\mathbf{F}$ and its integral, while those knobs shown in section (a) are used to set entries of $\mathbf{x}_0$ and other parameter values. There are two possible outputs: the plotter on the left, and the oscilloscope on the right. The equipment shown is mounted on a cart and can easily be taken into the classroom. All the examples of this paper can be simulated on the very modest computer shown in Figure 1.

AMS (MOS) subject classifications (1970). Primary 34-04; Secondary 98B99.

FIGURE 1

3. The scalar equation. Mathematics students soon see special cases of
the initial value problem of the form studied by Poincaré:

(3.1) $dy/dx = p(x, y)/q(x, y), \quad y(x_0) = y_0.$

Direct computer solutions of (3.1) can be impossible in a neighborhood of a
zero of $q(x, y)$. However, we can, in most cases, bypass this difficulty by considering the following orbit form in which the parameter t has been introduced:
For constant $k \neq 0$,

(3.2) $\dfrac{d}{dt}\begin{pmatrix} x \\ y \end{pmatrix} = k\begin{pmatrix} q(x, y) \\ p(x, y) \end{pmatrix}, \quad \begin{pmatrix} x(0) \\ y(0) \end{pmatrix} = \begin{pmatrix} x_0 \\ y_0 \end{pmatrix}.$

The orbits of (3.2) lie on integral curves of (3.1); the sign of k determines the
direction that the point $(x(t), y(t))$ travels along the orbit as t increases, and
its magnitude determines the speed.

A simple but rewarding example is the special case of (3.2): For constant

$k \neq 0$,

(3.3)
$$\frac{d}{dt}\binom{x}{y} = k\binom{x - x_1}{y - y_1}, \quad \binom{x(0)}{y(0)} = \binom{x_0}{y_0}$$

whose orbit is the straight line initiating at (x_0, y_0) and passing through (x_1, y_1).

As an application, consider the curve in 3-space

$$P(t) = \left((1 + 9\cos t)\cos t, (1 + 9\cos t)\sin t, \frac{3}{2}t \, \sin \frac{t}{2}\right).$$

The line segments $\overline{P(t_j)P(t_j + \pi)}$ for $0 \leqslant t_j \leqslant \pi$ and then projected onto the plane of the plotter and equation (3.3) with $k = -1$ is then used to construct these projected line segments as shown in Figure 2.

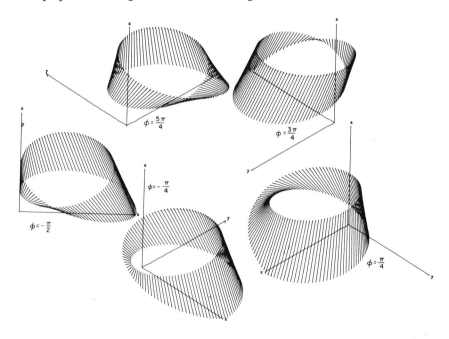

FIGURE 2

The projection mapping used here,

(3.4)
$$\binom{\bar{x}}{\bar{y}} = \begin{pmatrix} -\sin \phi & \cos \phi & 0 \\ -\sin \lambda \cos \phi & -\sin \lambda \sin \phi & \cos \lambda \end{pmatrix}\begin{pmatrix} x \\ y \\ z \end{pmatrix},$$

is itself a bit of applied linear algebra. In essence, this mapping sets the three-dimensional figure containing the point (x, y, z) above the plotter, and projects it orthogonally onto the plane of the plotter. Let $\mathbf{N}$ be a vector normal to the plane of the plotter. Then, in (3.4), ϕ is the angle that the projection of $\mathbf{N}$ onto the xy-coordinate plane makes with the positive x-axis, and λ is the angle $\mathbf{N}$ makes with the xy-plane (see equation (8) and Figure 3 of [3] with ϕ replaced by $\phi - \pi/2$).

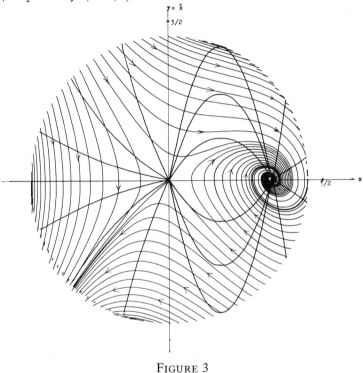

FIGURE 3

The equation

(3.5) $d^2x/dt^2 + \alpha\, dx/dt - x + x^2 = 0$

arises from a study of a steady shock wave that passes through a chain of particles connected by a nonlinear force [2]. If we let $y = dx/dt$, then (3.5) can be expressed as

(3.6) $\dfrac{d}{dt}\begin{pmatrix} x \\ y \end{pmatrix} = \begin{pmatrix} y \\ x - \alpha y - x^2 \end{pmatrix},$

which is of the form (3.2). Orbits of (3.6) are shown in Figure 3 with arrows indicating the direction in which $(x(t), y(t))$ traces out the orbit for increasing t. The curves without a direction indicated are isoclines.

If (x_0, y_0) is on the graph of

(3.7) $F(x, y) = 0,$

then it follows from elementary calculus that, for $k \neq 0$, the orbit

(3.8) $\dfrac{d}{dt}\begin{pmatrix} x \\ y \end{pmatrix} = k\begin{pmatrix} \partial F/\partial y \\ -\partial F/\partial x \end{pmatrix}, \quad \begin{pmatrix} x(0) \\ y(0) \end{pmatrix} = \begin{pmatrix} x_0 \\ y_0 \end{pmatrix}$

will lie on (3.7). For example, the conic section $(1 - \epsilon^2)x^2 + y^2 - 2\alpha x = C$, having ϵ as eccentricity, contains the orbit

$$\frac{d}{dt}\begin{pmatrix} x \\ y \end{pmatrix} = k\begin{pmatrix} y \\ -(1 - \epsilon^2)x + \alpha \end{pmatrix}, \quad \begin{pmatrix} x(0) \\ y(0) \end{pmatrix} = \begin{pmatrix} 1 \\ -2 \end{pmatrix}.$$

Figure 4 shows the plotter output of the analog solution for $\alpha = 1$, $k = \pm 1$, and the indicated eccentricities. This makes a nice classroom demonstration in that only one potentiometer sets ϵ.

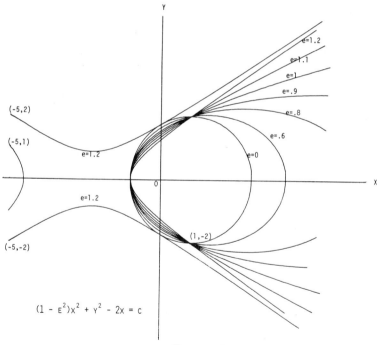

FIGURE 4

4. **Space curves.** Consider the n-dimensional curve defined as the simultaneous solution of the equations

(4.1) $\qquad F_j(x_1, x_2, \cdots, x_n) = 0, \qquad j = 1, 2, \cdots, n - 1,$

and assume that $\alpha = (\alpha_1, \alpha_2, \cdots, \alpha_n)$ is a point on this curve. It follows [4] that the orbit

(4.2)
$$
\frac{d}{dt}\begin{pmatrix} x_1 \\ x_2 \\ \cdot \\ \cdot \\ \cdot \\ x_n \end{pmatrix} = k \begin{pmatrix} \dfrac{\partial(F_1, F_2, \cdots, F_{n-1})}{\partial(x_2, x_3, \cdots, x_n)} \\[2mm] -\dfrac{\partial(F_1, F_2, \cdots, F_{n-1})}{\partial(x_1, x_3, \cdots, x_n)} \\[2mm] \cdots\cdots\cdots\cdots\cdots \\[2mm] (-1)^{n+1}\dfrac{\partial(F_1, F_2, \cdots, F_{n-1})}{\partial(x_1, x_2, \cdots, x_{n-1})} \end{pmatrix}, \qquad \begin{pmatrix} x_1(0) \\ x_2(0) \\ \cdot \\ \cdot \\ \cdot \\ x_n(0) \end{pmatrix} = \begin{pmatrix} \alpha_1 \\ \alpha_2 \\ \cdot \\ \cdot \\ \cdot \\ \alpha_n \end{pmatrix}
$$

for $k \neq 0$ will lie on (4.1).

An example for $n = 3$ is the curve

$$ x^2 - z + z^2 = 0, \qquad y - xz = 0, $$

which contains the orbit

$$
\frac{d}{dt}\begin{pmatrix} x \\ y \\ z \end{pmatrix} = \begin{pmatrix} 1 - 2z \\ z - 2z^2 + 2x^2 \\ 2x \end{pmatrix}, \qquad \begin{pmatrix} x(0) \\ y(0) \\ z(0) \end{pmatrix} = \begin{pmatrix} 0 \\ 0 \\ 0 \end{pmatrix}
$$

as shown in the darker curve in Figure 5. The lighter curves are the projection of this curve onto the xy-, xz-, and yz-planes, obtained by setting the coefficients of $z, y,$ and x respectively to zero in the matrix product (3.4).

To illustrate extrema of a function $F(x, y)$, we merely consider the z sections

$$ z - F(x, y) = 0, \qquad z - C = 0, $$

for which (4.2) becomes

(4.3)
$$
\frac{d}{dt}\begin{pmatrix} x \\ y \\ z \end{pmatrix} = k \begin{pmatrix} - \partial F/\partial y \\ \partial F/\partial x \\ 0 \end{pmatrix}.
$$

Figures 6 and 7 illustrate respectively the surfaces

$$ z = x^2 - 12y^2 + 4y^3 + 3y^4 \quad \text{and} \quad z = y^2 - x^3 - x^2 $$

as they are represented by their z sections, plotted using the analog realization of the corresponding forms of (4.3).

In a similar manner, we can use (4.2) to write equations similar to (4.3) for sections of more general algebraic surfaces. For example, Figures 8(a) and 8(b) show two views of the surface $z^3 - y^2 + x = 0$ as represented by its x sections, while Figures 9(a) and 9(b) show two views of the surface $y^2 - x^3 - x^2 - z^2 = 0$ as represented by its z sections.

5. The solution of an ordinary differential equation near a singular point. In [5], it is shown that the curve defined by

$$(5.1) \qquad (R(x)y') + Q(x, y) = 0, \qquad y(x_0) = y_0, \qquad y'(x_0) = y_0'$$

contains the projection of the orbit

$$(5.2) \qquad \frac{d}{dt}\begin{pmatrix} x \\ y \\ z \end{pmatrix} = k\begin{pmatrix} R(x) \\ z \\ -R(x)Q(x, y) \end{pmatrix}, \qquad \begin{pmatrix} x(0) \\ y(0) \\ z(0) \end{pmatrix} = \begin{pmatrix} x_0 \\ y_0 \\ -R(x_0)y_0' \end{pmatrix}$$

onto the xy-plane for $k \neq 0$. Say that $R(x_0) = 0$. Then x_0 may be a singular point of (5.1). Yet if $R(x)$ and $Q(x, y)$ are computable in a neighborhood of x_0, then (5.2) gives us a method for displaying solutions of (5.1) in a neighborhood of this singular point. As a special case, consider Bessel's equation

$$x^2 y'' + xy' + (x^2 - m^2)y = 0$$

having as a solution $J_m(x)$, the classical Bessel function of order m. We write the corresponding form of (5.2) in the quasilinear form

$$(5.3) \qquad \frac{d}{dt}\begin{pmatrix} x \\ y \\ z \end{pmatrix} = \begin{pmatrix} k & 0 & 0 \\ 0 & 0 & k \\ 0 & km^2 & 0 \end{pmatrix}\begin{pmatrix} x \\ y \\ z \end{pmatrix} + \begin{pmatrix} 0 \\ 0 \\ kx^2 y \end{pmatrix}, \qquad \begin{pmatrix} x(0) \\ y(0) \\ z(0) \end{pmatrix} = \begin{pmatrix} x_0 \\ J_m(x_0) \\ x_0 J_m'(x_0) \end{pmatrix}.$$

Figures 10(a)–(c) show orbits of the differential equation in (5.3) where $m = 1$, $x(0) = 16$, $y(0) = J_1(16)$, $z(0)$ takes on 22 different values, and $k = -1$ so that $x(t) \to 0$ as $t \to \infty$. The darker curve in Figure 10(c) is the graph of $J_1(x)$. In Figure 10(d), we see projections onto the xy-plane of another set of orbits of (5.3) with $m = 1$, the darker curve being the graph of the Neumann function $Y_1(x)$. These graphs illustrate a theorem (Theorem 4.1, p. 329 of [1]) stating that since the eigenvalues of the constant matrix on the right of (5.3) are $\lambda = k, \pm km$, and since we take $k < 0$, and $m = 1$, there is a two-dimensional manifold S in xyz-space containing the origin with the property that any orbit of (5.3) originating on S will tend to the origin as $t \to \infty$ and any solution originating off of X will be bounded away from the origin.

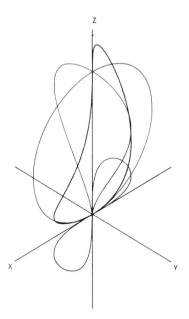

$$x^2 - z + z^2 = 0$$
$$y - xz = 0$$

FIGURE 5

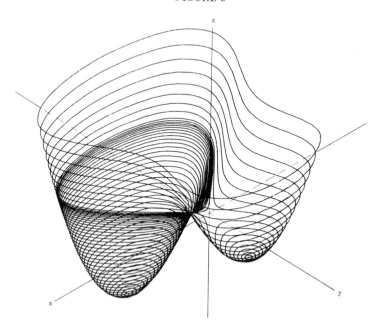

FIGURE 6

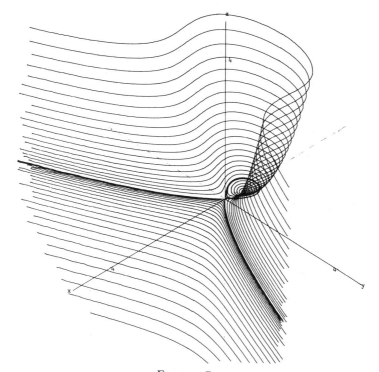

FIGURE 7

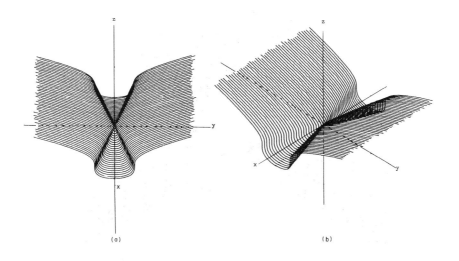

FIGURE 8

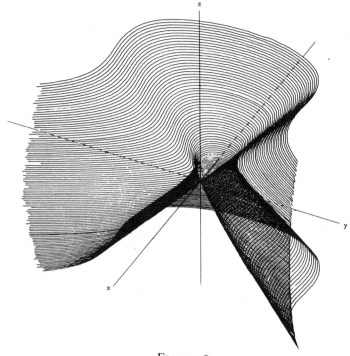

FIGURE 9a

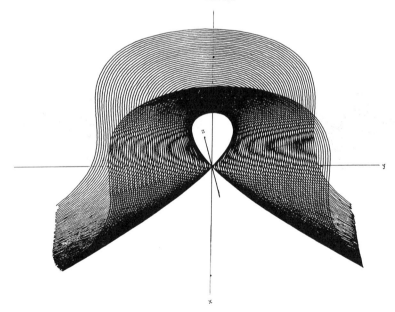

FIGURE 9b

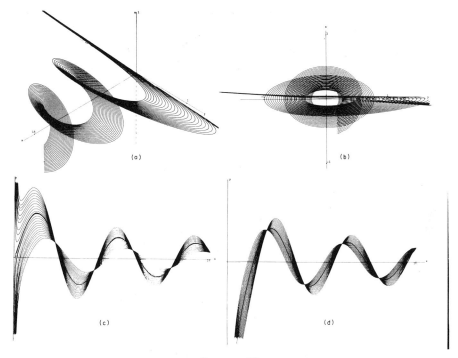

(a)   (b)

(c)   (d)

FIGURE 10

To generate polynomial functions of degree M on an analog computer can take up to M integrators and/or multipliers which could soon exceed the capabilities of many analog installations. In [6], the equation

(5.4) $[xD - (2N + m + 2)] \, [x^2D^2 + (x^2 - m^2)] \, y = 0$

is derived, having not only $J_m(x)$ as a solution, but the $(2N + 1)$st partial sum of the series expansion of $J_m(x)$,

$$y_{2N+1}(x) = \sum_{j=0}^{N} \frac{(-1)^j (x/2)^{2j+m}}{j! \, (j+1)!}.$$

Since $x = 0$ is a singular point of (5.4), techniques similar to those used to derive (5.2) were used to derive the equation

$$\frac{d}{dt}\begin{pmatrix} x \\ y \\ u \\ v \end{pmatrix} = k \begin{pmatrix} x \\ v \\ (2N + m + 2)u \\ u + (m^2 - x^2)y \end{pmatrix},$$

$k \neq 0$, whose orbits project onto solutions of (5.4) in the xy-plane. Figure 11 shows the solutions of (5.4) for $m = 1, J_1(x)$, and $y_{2N+1}(x)$ for $N = 3, 6,$ $\cdots$, 18. Thus in making the plot in Figure 11, we have generated polynomials of degree up to 37 using only four integrators and two multipliers.

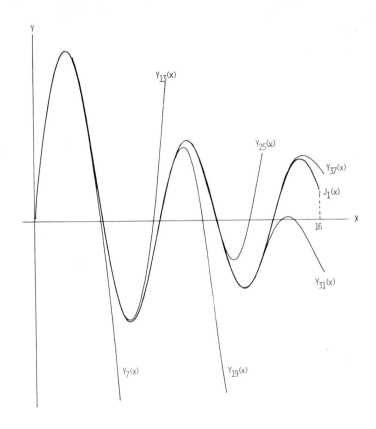

FIGURE 11

The latter two examples show not only how analog equipment can be used to illustrate mathematics, but also how we are able, with a knowledge of mathematics, to circumvent apparent limitations of the equipment.

6. Conclusion. The preceding examples do not exhaust the illustrative possibilities of a small analog computer. For example, a slight adaptation of the ideas presented above makes possible the illustration of ideas in complex analysis; Figure 12 shows level curves of the real and imaginary parts of e^z.

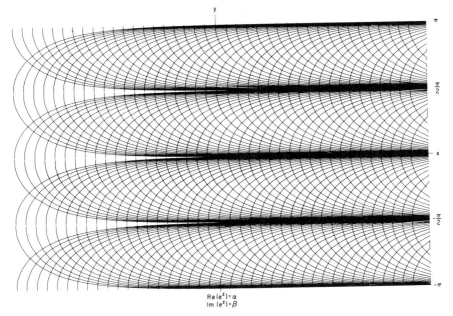

$$\text{Re}(e^z) = \alpha$$
$$\text{Im}(e^z) = \beta$$

FIGURE 12

In general, the analog computer has certain advantages over the digital computer when it comes to illustrating mathematical concepts. First is the hands-on capacity and immediate return. The programming is quite simple since one has essentially only addition, multiplication, raising to powers, their inverse operations, and integration. In programming, one goes directly to the mathematical structure of the concept. There are no additional complications such as difference approximations and questions concerning their convergence. At the start, the student sees the analog computer as an animated blackboard, but as time goes on he begins to see it as the electronic realization of many fascinating mathematical concepts.

REFERENCES

1. E. A. Coddington and N. Levinson, *Theory of ordinary differential equations*, McGraw-Hill, New York, 1955. MR 16, 1022.

2. G. E. Duval, R. Manvi and S. C. Lowell, *Steady shock profile in a one-dimensional lattice*, J. Appl. Phys. 40 (1969), 3755–3771.

3. T. A. Newton, *A mathematician uses the analog to illustrate mathematical concepts*, Analog/Hybrid Computer Educational Soc. Trans. 4 (1972), 183–193.

4. ———— , *On using the analog computer to illustrate space curves*, Proc. Second Annual Houston Conference on Circuits, Systems, and Computers, April 20–21, 1970, pp. 298–307.

5. ——— , *Some parametric techniques in the analog solution of ordinary differential equations*, IEEE Trans. Comput. (to appear).

6. ——— , *On using a differential equation to generate polynomials*, Amer. Math. Monthly (to appear).

WASHINGTON STATE UNIVERSITY

Proceedings of Symposia in Applied Mathematics
Volume 20
1974

AN INEXPENSIVE COMPUTER ASSIST IN TEACHING
LARGE ENROLLMENT MATHEMATICS COURSES

BY

EDWARD L. SPITZNAGEL, JR.

A problem common to many mathematics departments is providing good homework and examination procedures in large enrollment courses. If homework is required, the mass of assignments turned in encourage the copying of work and can be graded cursorily at best. If homework is not required, the students' motivation hinges on working a few unrealistically simple problems under intense examination pressures. As an alternative to these two states of affairs the author has developed a system of computer-generated problem sets containing realistically difficult problems to be worked by students at home. So far, in an experimental group of 23 students, the results have been very successful, and the system is now ready to be tried in a large enrollment course.

For concreteness this paper will discuss the method exactly as it was used in the experiment. However, in the course of the discussion it will become apparent that there is great flexibility in computer-generated problem sets–flexibility that can be used to advantage in adapting them to different kinds and levels of courses.

The computer is used to generate many problem sets, each with the same wording to the questions, but each one with different numbers inserted into the text of the questions. Two examples from the experimental course follow:

Compute estimates of the coefficients in the regression equation $E(Y) = A + BX$, given the following pairs of Y and X values:

Values of Y: 93, 74, 75, 57, 78.

Values of X: 91, 54, 72, 46, 58.

Suppose you are tested for diabetes and the test result is positive (indicating you may be diabetic). Suppose you know that 1.6% of the U.S. population is diabetic, that the test gives positive results 12.3% of the time when applied to the

---

AMS (MOS) subject classifications (1970). Primary 98B20, 98C05; Secondary 98A25.

175

population in general, and that the test applied to a diabetic has 96.7% chance of giving a positive result. What is your probability of being diabetic, given that your test is positive?

As these two examples indicate, the experimental course was one in probability and statistics. In fact, it was the difficulty of examining students on highly computational problems, exemplified by the regression problem above, that first led the author to consider computer-generated problem sets, each one with different data, that the students could work at home. It then became apparent that nearly all the problems on which the students were being tested could be handled the same way, since realistically complicated problems normally involve several different numbers that can be changed from one problem set to the next.

The students are allowed, an ample but fixed amount of time to work the problems. A student picking up his problem set at the first meeting that the sets are available must hand it in two class meetings (four or five days) hence. Any student may delay picking up his problem set until the second class meeting the sets are available (i. e., two or three days later) and then hand it in two class meetings following the time he has picked it up. A student presenting a serious medical or personal excuse is permitted to pick up his problem set as late as necessary and is given the same span of time to complete it.

The students are permitted to use textbooks, notes, any computational aids, and even to discuss the problems within limits that seem reasonable to them. By giving them such opportunities, the author hopes he is training them to do mathematics in the same surroundings they will be in after they graduate. Because the students are allowed to use so many resources, it is reasonable to make the problems realistically long, insert into them realistically complicated numbers, and to require the student to produce answers realistically close to the correct ones. The author uses a "rough slide-rule accuracy" criterion in scoring the problem sets: Any answer within 3% of the correct answer receives full credit, and any answer not within that limit receives no credit. The author believes that such an all-or-none grading scheme is also good preparation for the students' later use of mathematics. In grading time pressure examinations, instructors find it advantageous and reasonable to give liberal partial credit in case of wrong answers due only to arithmetic errors. However, if a professional mathematician causes a serious loss to his employer, his superiors do not ordinarily become more forgiving upon hearing that the loss was due to an arithmetic error rather than a conceptual mistake.

Since the student is graded only on his answers, all he needs to do is fill in the blank spaces provided on his problem set sheet and hand it in to be graded.

His results are then compared with a computer-generated answer sheet, and he receives his graded problem set at the next class meeting. Getting the graded problems back quickly, before the student forgets about his work, is a powerful aid in reinforcing learning, but even greater reinforcement is available through the repeatable feature of computer-generated problem sets.

Since pairs of problem sets differ completely in the numbers, any student wishing a second chance can be given a fresh problem set. Giving second chances seems realistic in terms of the way most mathematics students will do their work after graduating, but even if it were not, the author would advocate doing so, for they encourage students to learn from their failures as no ordinary examination system can: Before he attempts the second-chance problem set, the student finds it in his own best interest to locate and correct all errors on the first problem set. He can do so with full verification since by then he knows the correct answers to the first set. Different methods of scoring the second-chance work could be appropriate to different courses. The author requires the students to do only those problems they missed on their first sets and then averages the scores (that is, gives half credit for success the second time).

Some indications of the system's success are as follows: First, with the second-chance feature, nearly every student successfully learned to work nearly every problem. Second, when asked their preference, the students unanimously chose to continue with the system rather than revert to hour examinations—despite the fact that their scores on hour examinations the preceding semester had been slightly higher. Third, the correlation between the system and conventional testing was high, the sample correlation coefficient being 0.595. Besides these indications, some students volunteered that they enjoyed the opportunity to talk problems over with each other and that they seemed to be learning more by doing so. Despite the fact that the problem sets were mechanically produced and graded solely on the answers, the students seemed to find the system more human, more personal than conventional testing.

To use the system in a large enrollment low level course a few variations on the above description might be desirable. Frequent, easy problem sets would help the students keep pace with the material. Making the sets available at a centrally located distribution point instead of handling them out in class would save valuable time. One could still allow the option of picking up and returning a problem set late by having the students tear their sets off a serially numbered stack or stacks, so it becomes possible to tell which sets were taken the first day and which were not. Although it means more grading and bookkeeping work, the author definitely would recommend retention of the repeatable feature. For the first trial of the system with a very large number of students, the author would recommend using it in conjunction with at least one conventional examination, to guard against the one abuse the system is subject to: a good student working a poor student's problem set for him. No evidence of such cheating came to light in the author's experimental class, and, in fact, he thinks there are two reasons why cheating is unlikely. First, it requires more cooperation than the simple copying that can occur in examinations, for it requires a good student to make a conscious effort to do the work of another, poorer one. Second, a student who cheats often justifies his actions to himself by claiming that time-pressured examinations do not give him the chance to demonstrate his knowledge of the subject; with the computer-generated problem sets that excuse does not exist.

Almost any mathematics course in which computational skills are emphasized should be suited for use of computer-generated problem sets, especially algebra ("college algebra"), trigonometry, calculus, and ordinary differential equations courses. In each of these courses one has even greater freedom than that available in probability and statistics: One can alter polynomials and elementary functions to make the problem sets even less similar. One can also include graphing problems since it is possible to print rough graphs on the answer sheet, complete with upper and lower limit graphs.

Cost of the system is modest, especially in light of its benefits. In the author's experimental course, total cost was roughly $35, or $1.50 per student. In large enrollment courses the cost per student would be smaller yet, as the compilation time would be a smaller fraction of the total central processor time. Programming time is also low. The author, by no means a professional programmer, averaged approximately one hour per problem to produce both the text of the question and the routine to compute the answer.

Two other approaches to computer-generated problem sets have been reported. We mention them here to give a further indication of the versatility inherent in the method. One approach [1] consists in drawing questions at random from a large bank of (say, 300) questions, complete with multiple choice answers. No variations are made within the questions themselves. The problem sets thus produced are administered as timed examinations, but since every examination is different the students can be allowed choices of when to take their examinations and can be allowed to repeat them. Thus, some pressure is removed from the examination process even though the examinations are taken under time limits. An advantage of this system is its applicability to nonnumeric subjects such as history or foreign languages. One difficulty with it has been a tendency of some students to delay taking the examinations until some of the questions have become known by gossip.

A system using both a question bank and alteration of the contents of each question has been developed by Johnson [2], [3]. Johnson produces the first hour examination in freshman chemistry by drawing from a bank of 100 questions in which all of the following can be altered:

(1) Numbers.

(2) Elements and compounds.

(3) Location of the answer in a multiple-choice field.

Each examination consists of 20 multiple-choice questions with 5 possible answers per question. The examinations are given under time pressure, but the students are given five opportunities to take the examinations, with the highest score recorded as the grade. Again, time pressure is present but is lessened by the repeatable feature. Johnson reports that the average number of times a student takes the examination is close to four, just short of the maximum allowed. Appeal of the system can be judged from the fact that 95% of the students using it favored extending it to the second semester.

Johnson's system seems by far the most sophisticated one in existence. It probably is too sophisticated to be duplicated often; Johnson has indicated that roughly 1000 programmer hours went into its development. However, it can serve everyone as a valuable prototype, for it contains many innovations applicable to disciplines other than chemistry.

The author hopes the material in this paper has given the reader an idea of the power and usefulness of computer-generated problem sets. He also hopes the reader will consider ways they can be used to improve present examination procedures, especially in large enrollment courses, where the need for improvement seems greatest.

REFERENCES

1. *Proceedings of a Fourth Conference on Computers in Undergraduate Curricula*, University of Iowa, Iowa City, 1972, pp. 207–233.

2. K. J. Johnson, *Pitt's computer-generated repeatable chemistry exam*, Proc. Fourth Conference on Computers in the Undergraduate Curricula, University of Iowa, Iowa City, 1972, pp. 199–204.

3. K. J. Johnson and L. M. Epstein, *Pitt's computer-generated repeatable chemistry exam*, University of Pittsburgh, Pittsburgh, Pa., 1972.

WASHINGTON UNIVERSITY

Proceedings of Symposia in Applied Mathematics
Volume 20
1974

A NEW COMPUTER ORIENTED (ALGORITHMIC)
LINEAR ALGEBRA COURSE–PRELIMINARY REPORT

BY

ROBERT DUCHARME

In this paper I wish to discuss a new approach to the use of the computer as a teaching tool in a beginning course in linear algebra. The material being presented was developed by Professor Warren Stenberg of the University of Minnesota and the author.

The text itself includes material for at least two quarters of a beginning course in linear algebra, for undergraduates, with no prerequisite mathematics. The development of the materials is structured as follows. The linear transformation and its associated matrix are the basic elements. The fundamental operation on matrices is the column operation. Flowcharts are used exclusively in the development and presentation of all algorithms. This generalization makes the materials adaptable to all of the commonly used computer procedural languages.

Our feeling about computer mathematics is that oftentimes the computing is merely appended to the standard mathematical development. This "hanging on" of computing to the various topics often produces a collection of seemingly distinct algorithms with the result that the computing ideas compete with the mathematics rather than supplement it.

Our new approach unifies the main elements of a first course in that the central ideas are structured and developed around the construction of one principal algorithm, called the target algorithm, which unifies the course in a dramatic fashion. The material now presented illustrates the central ideas of the first two chapters, where this algorithm is developed.

As mentioned above, this linear algebra course begins with the concept of a linear transformation T from n-space into m-space. With very little effort on the part of the student, as well as the instructor, some topics which can immediately be "uncovered" include:

AMS (MOS) subject classifications (1970). Primary 15-01, 00-01.

For $T: R_n \to R_m$ with associated matrix A,

(1) find a basis for the range space of T;

(2) find a basis for the null space of T;

(3) when $T(x) = Ax = y$, determine whether a solution exists for the system;

(4) determine the inverse of A, when it exists.

We begin our development by assuming that a student entering a first course in linear algebra certainly has in his background the idea of solving a system of equations—be it by any of the methods taught. We start with this familiar concept and introduce only notation to get immediately to the idea of a linear transformation.

For example, we view a system such as

(1)
$$
\begin{aligned}
1x_1 - 6x_2 + 1x_3 + 4x_4 &= 3, \\
-2x_1 + 14x_2 + 1x_3 - 9x_4 &= -5, \\
2x_1 - 6x_2 + 11x_3 + 5x_4 &= 9,
\end{aligned}
$$

in standard matrix notation as

(2)
$$
T(x) = Ax = \begin{pmatrix} 1 & -6 & 1 & 4 \\ -2 & 14 & 1 & -9 \\ 2 & -6 & 11 & 5 \end{pmatrix} \begin{pmatrix} x_1 \\ x_2 \\ x_3 \\ x_4 \end{pmatrix} = \begin{pmatrix} 3 \\ -5 \\ 9 \end{pmatrix} = y.
$$

After observing that for the standard basis vectors $e^{(i)}$, for $i = 1, 2, 3, 4$,

(3)
$$
T(e^{(i)}) = \text{column}_i(A),
$$

we construct the "natural double matrix representation" of the linear transformation T. As one can see in (4), we have stacked the identity matrix I on top of the matrix A of T:

(4)
$$
D = \left(\frac{I}{A} \right) = \begin{pmatrix} 1 & 0 & 0 & 0 \\ 0 & 1 & 0 & 0 \\ 0 & 0 & 1 & 0 \\ 0 & 0 & 0 & 1 \\ \hline 1 & -6 & 1 & 4 \\ -2 & 14 & 1 & -9 \\ 2 & -6 & 11 & 5 \end{pmatrix}.
$$

The horizontal line in D is used to separate the lower and upper portions of D. On occasion we will refer to D as a single matrix while at other times we will discuss its "top" and "bottom" portions. Similarly, we will have occasion to consider the column vectors of D either as $(m + n)$-vectors or as m-vectors and n-vectors.

From our observation in (3) we see that for each column vector of D, the image of the top portion of the vector, $T(e^{(i)})$, for $i = 1, 2, 3, 4$, is sitting directly below it as $\text{column}_i(A)$.

Once the standard elementary column operations on matrices have been introduced, that is, multiply a column by a nonzero scalar, multiply a column by a scalar and add to a second column, and interchange two columns, we prove the

THEOREM. *Column operations on a (natural) double matrix representation* D *of a transformation* T *preserve the following properties of* D:

(i) *The resultant column vectors in the top portion of* D *still form a basis for the domain* R_n.

(ii) *Each column vector in the top portion still maps, respectively, into the column vector directly below it.*

With this development we return to the example in (4) and proceed to reduce its lower portion to echelon form.

$$
(5)\quad
\begin{pmatrix}
1 & 0 & 0 & 0 \\
0 & 1 & 0 & 0 \\
0 & 0 & 1 & 0 \\
0 & 0 & 0 & 1 \\
\hline
1 & -6 & 1 & 4 \\
-2 & 14 & 1 & -9 \\
2 & -6 & 11 & 5
\end{pmatrix}
\rightarrow
\begin{pmatrix}
1 & 6 & -1 & -4 \\
0 & 1 & 0 & 0 \\
0 & 0 & 1 & 0 \\
0 & 0 & 0 & 1 \\
\hline
1 & 0 & 0 & 0 \\
-2 & 2 & 3 & -1 \\
2 & 6 & 9 & -3
\end{pmatrix}
\rightarrow
\begin{pmatrix}
1 & 3 & -1 & -4 \\
0 & 1/2 & 0 & 0 \\
0 & 0 & 1 & 0 \\
0 & 0 & 0 & 1 \\
\hline
1 & 0 & 0 & 0 \\
-2 & 1 & 3 & -1 \\
2 & 3 & 9 & -3
\end{pmatrix}
\rightarrow
\begin{pmatrix}
7 & 3 & -10 & -1 \\
1 & 1/2 & -3/2 & 1/2 \\
0 & 0 & 1 & 0 \\
0 & 0 & 0 & 1 \\
\hline
1 & 0 & 0 & 0 \\
0 & 1 & 0 & 0 \\
8 & 3 & 0 & 0
\end{pmatrix}.
$$

We partition the final double matrix in (5) by drawing a vertical line which separates the columns of zeroes in the lower right quadrant:

$$
(6)\quad
\left(
\begin{array}{cc|cc}
7 & 3 & -10 & -1 \\
1 & 1/2 & -3/2 & 1/2 \\
0 & 0 & 1 & 0 \\
0 & 0 & 0 & 1 \\
\hline
1 & 0 & 0 & 0 \\
0 & 1 & 0 & 0 \\
8 & 3 & 0 & 0
\end{array}
\right)
=
\left(
\begin{array}{c|c}
Q & N \\
\hline
R & Z
\end{array}
\right).
$$

With the partitioning complete we are now ready to study the implications of this final target form. From our theorem we know that the four vectors in the upper portion of (6) form a basis for R_4. Thus, for $x \in R_4$,

$$(7) \quad x = c_1 \cdot \begin{pmatrix} 7 \\ 1 \\ 0 \\ 0 \end{pmatrix} + c_2 \cdot \begin{pmatrix} 3 \\ 1/2 \\ 0 \\ 0 \end{pmatrix} + c_3 \cdot \begin{pmatrix} -10 \\ -3/2 \\ 1 \\ 0 \end{pmatrix} + c_4 \cdot \begin{pmatrix} -1 \\ 1/2 \\ 0 \\ 1 \end{pmatrix}.$$

Its image under T can then be represented as

$$T(x) = c_1 \cdot T\begin{pmatrix} 7 \\ 1 \\ 0 \\ 0 \end{pmatrix} + c_2 \cdot T\begin{pmatrix} 3 \\ 1/2 \\ 0 \\ 0 \end{pmatrix} + c_3 \cdot T\begin{pmatrix} -10 \\ -3/2 \\ 1 \\ 0 \end{pmatrix} + c_4 \cdot T\begin{pmatrix} -1 \\ 1/2 \\ 0 \\ 1 \end{pmatrix}$$

$$= c_1 \begin{pmatrix} 1 \\ 0 \\ 8 \end{pmatrix} + c_2 \begin{pmatrix} 0 \\ 1 \\ 3 \end{pmatrix} + c_3 \begin{pmatrix} 0 \\ 0 \\ 0 \end{pmatrix} + c_4 \begin{pmatrix} 0 \\ 0 \\ 0 \end{pmatrix}$$

$$(8) \quad = c_1 \begin{pmatrix} 1 \\ 0 \\ 8 \end{pmatrix} + c_2 \begin{pmatrix} 0 \\ 1 \\ 3 \end{pmatrix}$$

$$(9) \quad = \begin{pmatrix} c_1 \\ c_2 \\ 8c_1 + 3c_2 \end{pmatrix}.$$

We see from (8) that all image vectors are spanned by the vectors

$$\begin{pmatrix} 1 \\ 0 \\ 8 \end{pmatrix} \quad \text{and} \quad \begin{pmatrix} 0 \\ 1 \\ 3 \end{pmatrix}.$$

Thus the vectors in the lower left quadrant of (6), that is, in quadrant R, form a basis for the range space of T.

Moreover, if x is in the null space of T, we have

$$T(x) = \begin{pmatrix} 0 \\ 0 \\ 0 \end{pmatrix} = \begin{pmatrix} c_1 \\ c_2 \\ 8c_1 + 3c_2 \end{pmatrix}$$

from (9). Thus $c_1 = c_2 = 0$ and x must be of the form

$$x = 0 \begin{pmatrix} 7 \\ 1 \\ 0 \\ 0 \end{pmatrix} + 0 \begin{pmatrix} 3 \\ 1/2 \\ 0 \\ 0 \end{pmatrix} + c_3 \begin{pmatrix} -10 \\ -3/2 \\ 1 \\ 0 \end{pmatrix} + c_4 \begin{pmatrix} -1 \\ 1/2 \\ 0 \\ 1 \end{pmatrix}$$

$$= c_3 \begin{pmatrix} -10 \\ -3/2 \\ 1 \\ 0 \end{pmatrix} + c_4 \begin{pmatrix} -1 \\ 1/2 \\ 0 \\ 1 \end{pmatrix}$$

where we see that the vectors

$$\begin{pmatrix} -10 \\ -3/2 \\ 1 \\ 0 \end{pmatrix} \text{ and } \begin{pmatrix} -1 \\ 1/2 \\ 0 \\ 1 \end{pmatrix}$$

which are found in the upper right quadrant N of the target matrix in (6) form a basis for the null space of T.

Thus we can, by inspection of the target matrix

$$\left(\begin{array}{c|c} Q & N \\ \hline R & Z \end{array} \right),$$

conclude the following:

(i) The column vectors of quadrant R form a basis for the range of T.

(ii) The column vectors of quadrant N form a basis for the null space of T.

(iii) The column vectors of quadrant Z are all zero.

(iv) The column vectors of quadrant Q are domain vectors whose images form a basis for the range of T.

We now investigate how our target matrix helps to solve a system of equations as in (1). Our students consider the problem as being of the form: Is the vector $(3 \ -5 \ 9)^T$ an image vector? That is, is it of the form

$$\begin{pmatrix} c_1 \\ c_2 \\ 8c_1 + 3c_2 \end{pmatrix}.$$

Since it is, the general solution from (7) with $c_1 = 3$ and $c_2 = -5$ is

$$x = 3 \begin{pmatrix} 7 \\ 1 \\ 0 \\ 0 \end{pmatrix} + (-5) \begin{pmatrix} 3 \\ 1/2 \\ 0 \\ 0 \end{pmatrix} + c_3 \begin{pmatrix} -10 \\ -3/2 \\ 1 \\ 0 \end{pmatrix} + c_4 \begin{pmatrix} -1 \\ 1/2 \\ 0 \\ 1 \end{pmatrix}.$$

And in general, for the system $T(x) = (y_1 \ y_2 \ y_3)^T$ if $8y_1 + 3y_2 = y_3$, our solution is

$$x = y_1 \begin{pmatrix} 7 \\ 1 \\ 0 \\ 0 \end{pmatrix} + y_2 \begin{pmatrix} 3 \\ 1/2 \\ 0 \\ 0 \end{pmatrix} + c_3 \begin{pmatrix} -10 \\ -3/2 \\ 1 \\ 0 \end{pmatrix} + c_4 \begin{pmatrix} -1 \\ 1/2 \\ 0 \\ 1 \end{pmatrix}.$$

And, of course, for the homogeneous system, the solution is

$$x = c_3 \begin{pmatrix} -10 \\ -3/2 \\ 1 \\ 0 \end{pmatrix} + c_4 \begin{pmatrix} -1 \\ 1/2 \\ 0 \\ 1 \end{pmatrix}.$$

If the matrix of the transformation has an inverse, we find that it is produced automatically in the target matrix reduction. Consider, for example, the
matrix

$$A = \begin{pmatrix} 1 & 2 & -2 \\ 3 & 7 & -4 \\ -1 & 5 & 15 \end{pmatrix}.$$

Applying the reduction algorithm to the natural double matrix we obtain

$$\left(\frac{I}{A} \right) = \begin{pmatrix} 1 & 0 & 0 \\ 0 & 1 & 0 \\ 0 & 0 & 1 \\ 1 & 2 & -2 \\ 3 & 7 & -4 \\ -1 & 5 & 15 \end{pmatrix} \rightarrow \begin{pmatrix} -125 & 40 & -6 \\ 41 & -13 & 2 \\ -22 & 7 & -1 \\ 1 & 0 & 0 \\ 0 & 1 & 0 \\ 0 & 0 & 1 \end{pmatrix} = \left(\frac{A^{-1}}{I} \right).$$

However, when the student attempts to partition the final target matrix, he ob-
serves that there are no columns of zeroes. The null space is then trivial. Moreover, the inverse of A is in the upper portion of the final target matrix.

The target matrix algorithm is easily flowcharted. Note that frames 9 and
10, respectively, are subprogram procedures which normalize the pivot element
and then use multiples of it to blast the pivot row elements to zero. See Figure 1.

In Figure 2 we see a modification of the flowchart to take into account the
possibility of a zero pivot element. Frames 5 and 8 are subprogram procedures
which perform the partial pivoting. Finally, in Figure 3 we add the few remaining
steps necessary to solve for the system of equations.

Thus, at this point, a single algorithm has been constructed to deal with
several of the basic topics in linear algebra. Moreover, the same algorithm can now
be used to solve further problems of linear algebra. For example, if C is a set
(finite) of vectors, the target algorithm can be used in the following way:

(i) Find a basis for the space spanned by C. (The columns of quadrant
R furnish this.)

(ii) Determine if C is a linearly independent set. (Check if quadrant Z
is empty.)

(iii) Find a nontrivial linear combination summing to zero if C is linearly
dependent. (If Z is not empty, use the components of any column vector of
quadrant N as coefficients.)

Target Matrix Flowchart

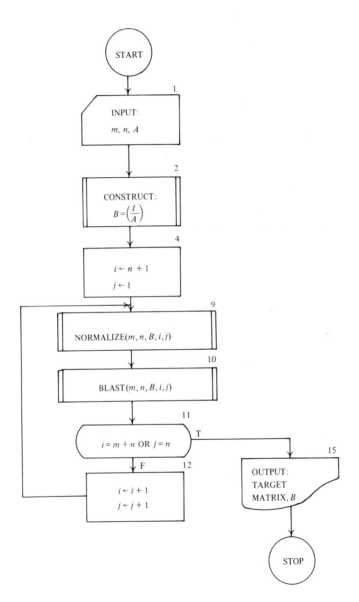

Figure 1

Target Matrix Flowchart with Partial Pivoting

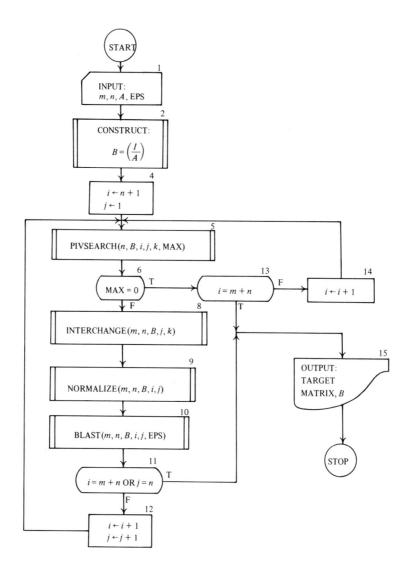

Figure 2

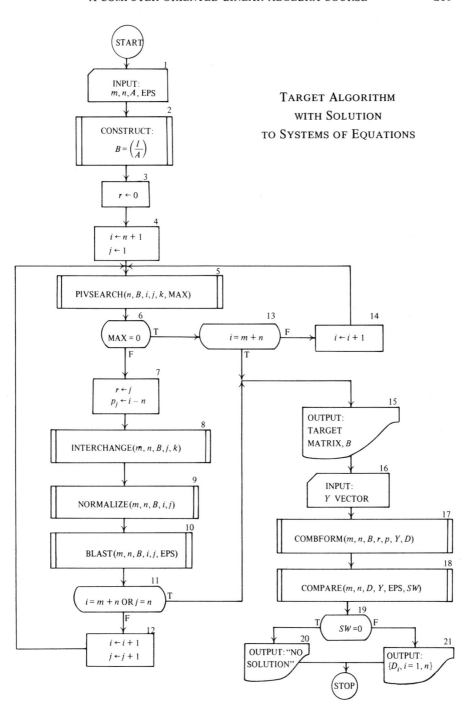

TARGET ALGORITHM
WITH SOLUTION
TO SYSTEMS OF EQUATIONS

FIGURE 3

(iv) Find the dimension of the space spanned by C. (It is the number of columns of quadrant R.)

(v) Find the orthogonal complement of C. (Construct the matrix A using as rows the vectors of C. Apply the target algorithm; the column vectors of quadrant N form a basis for $C^\perp$.)

(vi) Find among the vectors of C a basis for $[C]$. (Construct the matrix A using as rows the vectors of C. Pivot rows in the target matrix are basis rows in A.)

(vii) Find the determinant of A. (Introduce into the algorithm the three or four additional steps necessary to keep track of those column operations that affect the value of the determinant; print out the final value.)

There are more applications of this target algorithm in the text.

In summary, we have attempted to structure a course in elementary linear algebra which incorporates a spiral approach around a single, central algorithm whose construction and development are a part of the mathematics itself and not merely an appendage.

Other algorithms developed in the text are used for the characteristic polynomial, for eigenvalues of a symmetric matrix, and for the Jordan canonical form. The text was used in 1973–1974 at both the University of Minnesota and at Baldwin-Wallace College. More information can be obtained from the authors.

BALDWIN-WALLACE COLLEGE

Proceedings of Symposia in Applied Mathematics
Volume 20
1974

COMPUTER SUPPLEMENTED BUSINESS ORIENTED MATHEMATICS

BY

KENNETH L. HANKERSON AND GENE A. KEMPER

An important function of a mathematics department is to provide service courses. In recent years the clientele for such courses has been expanding beyond the usual engineering and "hard" science disciplines to include the "soft" sciences and non-sciences. While it is unrealistic for a mathematics department to offer service courses for each discipline requesting such courses it is equally unrealistic to expect such disciplines to require their students to take courses oriented to engineering, "hard" science, or mathematics students.

The Department of Mathematics at the University of North Dakota has initiated a three-phase development of four three-semester credit hour courses oriented toward the "soft" sciences and non-science students. The development of the first of the four courses has evolved during the past three years and is essentially complete. The use of a computer was an integral part of the development of this course and that development is the topic of this paper.

Since the initial demand for an appropriate service course came from the College of Business and Public Administration, the illustrations and problems in the course are presently oriented toward business applications. The course is at the sophomore level and does not have a college level prerequisite.

The computer used in the initial offering of the course was actually a programmable calculator, specifically, an Olivetti-Underwood Programma 101. The 101 uses an assembler type language and has a capacity for 48 instructions and 10 storage registers or up to 120 instructions at the expense of some of the storage capacity. About three hours of class time were spent for programming instruction. The programming language proved to be a handicap in that not only was it somewhat tedious to learn, but once a program had been completed the code did not look like the mathematics involved. Moreover, because of severe

AMS (MOS) subject classifications (1970). Primary 90-01, 98A20, 98B20.

storage limitations it was not possible to conveniently consider many realistic examples. As a result few programs were required of the students.

The second offering of the course was modified to the extent that the instructor provided several programs to be utilized by the students. The students were still taught to program the 101. However, by providing programs to the students they were able to spend more time learning mathematical methods and less time coding. Another severe handicap of the particular 101 used was that if the student made an error while programming the computer he would have to start over. (Apparently some models of the 101 have overcome this difficulty and an error can be corrected by an appropriate sequence of instructions.)

Between the second and fourth offering of this course the University acquired a PDP-8 time-sharing system. It was really access to this system that "jelled" the course. The time-sharing system is small in that it consists of 8K of memory with five user partitions. Nevertheless, such capacity was not a real hindrance to the course. Each of the ASR-33 teletypes has a paper tape reader and punch. BASIC is the only language available on the system. Two one-hour periods were used for programming instruction with slight additional instruction provided as needed. Not only is BASIC easily learned, but the code looks like the mathematics it represents. This is an important learning factor that should not be overlooked. Moreover, error diagnostics are given, programs are easily modified, less time is spent on coding, more realistic problems can be considered, and student motivation rose considerably—as did the instructor's motivation.

In the following brief outline of Mathematics 203, Mathematical Methods I, note that trigonometric functions are not included, but only polynomial, exponential, and logarithmic functions or combinations of these. Every topic required at least one program. Some of the topics consisted of a single example and a similar problem.

Outline

1. Definition of function and corresponding notation with emphasis on linear and quadratic functions.

2. Graphing.

3. Systems of linear equations with emphasis on the pivot method of solution as later utilized in the Gauss-Jordan algorithm as well as in the simplex method.

4. Absolute value and linear inequalities with emphasis on systems of linear inequalities as appear later in linear programming.

(*)   5. Matrix algebra with emphasis on the Gauss-Jordan algorithm as applied to the solution of linear systems, the inversion of matrices and the evaluation of determinants.

6. Markov processes. (A very brief introduction as an example of matrix applications.)

(*)   7. Linear programming using the simplex method and including elementary post-optimal analysis.

8. Linear integer programming. (A very brief introduction which required the students to write a program using a "brute force" method applied to an elementary example which illustrated that one should not solve the noninteger problem and then round the results.)

9. Exponents and logarithms. (Computational manipulation of logarithms was not emphasized but rather consideration was given to properties.)

10. Differential calculus with emphasis on the derivative as a rate of change. Differentiation rules were considered, but only formulas pertaining to polynomials, exponentials, and logarithms were considered. Partial differentiation was included.

11. Newton's method for solving a single equation.

12. Maximum-minimum applications with emphasis on supply-demand and cost-revenue-profit considerations.

13. Nonlinear integer programming with linear constraints. (A very brief introduction requiring the students to use a "brute force" method for investigating an inventory model.)

14. Linear, quadratic and exponential least squares.

15. Integral calculus via the Riemann sum. The Fundamental Theorem of Calculus.

Programs for those topics in the outline marked with an asterisk were furnished by the instructor. The programs are available to the student in the form of a University of North Dakota Computer Center Special Report No. 46 entitled *Gauss-Jordan elimination, linear programming and quadratic least squares for the* PDP-8. Each method is documented according to the categories: problem, program logic, program listing, mathematical restrictions, program restrictions, computer restrictions, program usage and examples. The elementary nature of this documentation makes it readily understandable to the student.

In order to most effectively utilize the programming efforts of the students a program assigned to be prepared by the students will frequently be used in a later topic. For example the students are required to write a program for finding a real zero of a real cubic polynomial using Newton's method. This program will be used later to determine the maximum profit of a cost-revenue-profit model.

The typical level of assigned problems is illustrated by the example in Appendix A.

Phase two of the development was initiated during the 1973-74 academic year. It consists of two courses, Mathematical Methods II and Principles of Statistical Sampling. The course description of all three courses is presented in Appendix B. Phase three, which consists of the development of applied statistics course, is not yet scheduled for implementation.

In summary, the existing course was developed to fill a specific need. Its evolution was greatly influenced by the use of a computer and by the particular computer used. The use of the time-sharing capability makes possible the meaningful and interesting, even though at times brief, introduction to several important topics.

APPENDIX A. INVENTORY CONTROL (TWO PRODUCT)

C = cost (in dollars) of inventorying q_i units of item i, $i = 1, 2$.

k_i = purchase price of one unit of item i.

c_2 = cost of placing an order.

c_3 = cost of holding one unit in inventory for the time period (expressed as percentage of purchase price).

D_i = demand for item i.

Model.

$$C = k_1 D_1 + k_2 D_2 + \frac{c_2 D_1}{q_1} + \frac{c_2 D_2}{q_2} + \frac{c_3 k_2}{2} q_1 + \frac{c_3 k_2}{2} q_2, \qquad q_1, q_2 > 0.$$

Problem 1. The *economic order quantity* is the values of q_1 and q_2, denoted $\bar{q}_1$ and $\bar{q}_2$, that minimize C. Find $\bar{q}_1$ and $\bar{q}_2$ if $D_1 = 2500$, $D_2 = 2000$, $k_1 = \$40$, $k_2 = \$12.50$, $c_2 = \$4$ and $c_3 = 10\%$. (The students solved this problem by using partial derivatives.)

Problem 2. Suppose the warehouse holds only 100 items. Then $\bar{q}_1$ and $\bar{q}_2$ need to be determined to satisfy the constraint $0 < q_1 + q_2 \leqslant 100$. (The method of Lagrange multipliers was discussed but the corresponding problem was not solved.)

Problem 3. A BASIC program was written by the students to solve the following nonlinear integer programming problem by enumerating all the feasible points and choosing the one that minimized C:

Minimize C

subject to $0 < q_1 + q_2 \leqslant 100,$ q_1, q_2 positive integers.

APPENDIX B. COURSE DESCRIPTION

Math 203. Mathematical Methods I (3 semester credits)

Prerequisites. Three semesters of high school algebra. (This is the same as the prerequisite for College Algebra, Math 103.)

Content. Students will be instructed in programming the PDP-8 time-sharing computer using the BASIC language. This implies the topics considered in 203 are approached from the viewpoint of providing useful tools for "real world" applications that are compatible with techniques utilizing a computer to obtain a solution. These topics include systems of linear equations and inequalities, introduction to vectors and matrices, linear programming, quadratic functions with emphasis on graphing and extrema, exponential and logarithmic functions, Newton's method for solving a single equation and topics in differential and integral calculus with appropriate applications.

Math 204. Mathematical Methods II (3 semester credits)

Prerequisites. Math 203 or consent of instructor.

Content. The course content is designed to provide a natural extension of the methods studied in Math 203 and to provide an introduction to concepts of probability and statistics needed in Math 324. The course includes further topics in differential and integral calculus with applications (including least squares curve fitting and partial differentiation), elementary probability and topics from elementary statistics. As in Math 203, the students will be required to use the PDP-8 time-sharing computer.

Math 324. Principles of Statistical Sampling (3 semester credits)

Prerequisites. Math 204 or consent of instructor; or a noncalculus elementary statistics course such as Biology 470, Economics 210, Psychology 241, Sociology 426, etc.

Content. Students will determine how and when to use various sample survey techniques by studying problems from different fields. The PDP-8 time-sharing computer will be used for computations. Students will actually perform sample surveys related to their field.

Included in this course will be a review of necessary concepts from probability and statistics, types of data-gathering methods (including questionnaire construction), construction and analysis of most common sampling schemes (including simple, stratified, cluster and 2-stage random sampling), ratio estimation and wildlife sampling.

UNIVERSITY OF NORTH DAKOTA

INDEXES

AUTHOR INDEX

SUBJECT INDEX